VOLUME 2: HOW TO PERFORM CONTINUOUS SAMPLING

SECOND EDITION

The ASQC Basic References in Quality Control: Statistical Techniques
Edward F. Mykytka, Ph.D., Editor

VOLUME 2: HOW TO PERFORM CONTINUOUS SAMPLING

Second Edition

Kenneth S. Stephens

Volume 2: How to Perform Continuous Sampling, Second Edition

Kenneth S. Stephens

Library of Congress Cataloging-in Publication Data

Stephens, Kenneth S.
   How to perform continuous sampling

   (The ASQC basic references in quality control; v. 2)
   Includes bibliographical references (p. 38–42).
   1. Acceptance sampling. I. Title. II. Series.
TS156.4.S74 1986   658.5'62'015195   92-110814
ISBN 0-87389-012-4 (v. 1)
ISBN 0-87389-330-1 (v. 2)

© 1995 by ASQC

All rights reserved. No part of this book may be reproduced in any form or by any means, electronic, mechanical, photocopying, recording, or otherwise, without the prior written permission of the publisher.

10 9 8 7 6 5 4 3 2 1

ISBN 0-87389-330-1

ASQC Mission: To facilitate continuous improvement and increase customer satisfaction by identifying, communicating, and promoting the use of quality principles, concepts, and technologies; and thereby be recognized throughout the world as the leading authority on, and champion for, quality.

For a free copy of the ASQC Quality Press Publications Catalog, including ASQC membership information, call 800-248-1946.

Printed in the United States of America

Printed on acid-free recycled paper

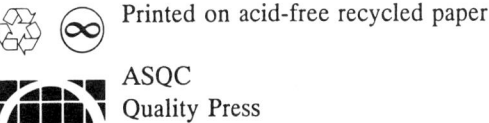

ASQC
Quality Press
611 East Wisconsin Avenue
Milwaukee, Wisconsin 53202

The ASQC Basic References in Quality Control: Statistical Techniques is a continuing literature project of ASQC's Statistics Division. Its aim is to survey topics in statistical quality control in a practically usable, "how-to" form in order to provide the quality practitioner with specific, ready-to-use tools for conducting statistical analyses in support of the quality improvement process.

Suggestions regarding subject matter content and format of the booklets are welcome and will be considered for future editions and revisions. Such suggestions should be sent to the series editor(s).

Volumes Published
Volume 1: How to Analyze Data with Simple Plots (W. Nelson)
Volume 2: How to Perform Continuous Sampling (K. S. Stephens)
Volume 3: How to Test Normality and Other Distributional Assumptions
  (S. S. Shapiro)
Volume 4: How to Perform Skip-Lot and Chain Sampling (K. S. Stephens)
Volume 5: How to Run Mixture Experiments for Product Quality (J. A. Cornell)
Volume 6: How to Analyze Reliability Data (W. Nelson)
Volume 7: How and When to Perform Bayesian Acceptance Sampling (T. W. Calvin)
Volume 8: How to Apply Response Surface Methodology (J. Cornell)
Volume 9: How to Use Regression Analysis in Quality Control (D. C. Crocker)
Volume 10: How to Plan an Accelerated Life Test—Some Practical Guidelines
  (W. Meeker and G. Hahn)
Volume 11: How to Perform Statistical Tolerance Analysis (N. Cox)
Volume 12: How to Choose the Proper Sample Size (G. G. Brush)
Volume 13: How to Use Sequential Statistical Methods (T. P. McWilliams)
Volume 14: How to Construct Fractional Factorial Experiments (R. F. Gunst and R. L. Mason)
Volume 15: How to Determine Sample Size and Estimate Failure Rate in Life Testing (E. C. Moura)
Volume 16: How to Detect and Handle Outliers (B. Iglewicz and D. C. Hoaglin)

Editorial Review Board

Nancy E. Baxter-Belunis
Thomas J. Lipton, Inc.

William M. Mead
Babcock & Wilcox

Gerald J. Hahn
General Electric Company

Annabeth Propst
Quality Transformation Services

Lynne B. Hare
Thomas J. Lipton, Inc.

Samuel S. Shapiro
Florida International University

Stuart J. Janis
3M Company

Harrison M. Wadsworth
Georgia Institute of Technology

Norman L. Johnson
University of North Carolina

William H. Woodall
University of Alabama

Saeed Maghsoodloo
Auburn University

Hassan Zahedi
Florida International University

Michael J. Mazu
Alcoa Corporation

The author wishes to thank the
Series' Review Board, series' editor,
the editorial and production staff
of ASQC Quality Press, and
Thomas M. Kubiak for their assistance
with the current revision.

CONTENTS

	Page
FOREWORD TO THE SECOND EDITION	xi
FOREWORD TO THE FIRST EDITION	xiii
ABSTRACT	1

1.0	BRIEF INTRODUCTION TO ACCEPTANCE SAMPLING PLANS AND CONCEPTS	1
1.1	BASIC ELEMENTS OF A SAMPLING PLAN AND ACCEPTANCE SAMPLING	1
1.2	EVALUATION OF THE PERFORMANCE OF SAMPLING PLANS—OC CURVE	2
1.3	QUALITY LEVEL INDEXING TO CLASSIFY SAMPLING PLANS—AQL, AOQL, LQL	2
1.4	TYPES OF SAMPLING PLANS	3
	1.4.1 ATTRIBUTES AND VARIABLES DATA AND SAMPLING PLANS	3
	1.4.2 SINGLE, DOUBLE, MULTIPLE, AND SEQUENTIAL SAMPLING PLANS	4
	1.4.3 LOT-BY-LOT SAMPLING, CONTINUOUS SAMPLING, AND CUMULATIVE SAMPLING	4
	1.4.4 SYSTEMS OF SAMPLING PLANS—TABLES AND PROCEDURES	4
1.5	ESSENTIAL ITEMS FOR THE PREPARATION AND DEVELOPMENT OF SAMPLING PLANS	4
	1.5.1 PRELIMINARY QUALITY AND TEST INFORMATION	5
	1.5.2 REFINED QUALITY AND ACCEPTABILITY INFORMATION	5
	1.5.3 DETERMINING SAMPLING METHODS AND SAMPLING PLANS	5
1.6	ARE SAMPLING PLANS PASSÉ?	6
2.0	CONTINUOUS SAMPLING PLANS (CSP)	9
2.1	BASIC PARAMETERS, OPERATION, EVALUATION, AND APPLICATION (CSP-1)	9
2.2	ADDITIONAL CONTINUOUS SAMPLING PLANS	13
	2.2.1 CSP-2 AND CSP-3	14
	2.2.2 CSP-A	17
	2.2.3 CSP-M	17
	2.2.4 CSP-T	19
	2.2.5 CSP-F	19
	2.2.6 CSP-V	19
	2.2.7 CSP-R	22
	2.2.8 CSP-C	22

2.3	NOMOGRAPHS AND THEIR USE	24
2.4	SAMPLING TABLES AND THEIR USE	34
	2.4.1 CSP-1	34
	2.4.2 CSP-2	35
	2.4.3 CSP-A	35
	2.4.4 CSP-M	36
	2.4.5 CSP-T	37
	2.4.6 CSP-F	37
	2.4.7 CSP-V	37
2.5	SUMMARY	38
APPENDIX A: DETERMINATION OF OPERATING CHARACTERISTICS BY MARKOV CHAINS		41
APPENDIX B: NOTE ON CSP-1 AND SkSP-1, EQUATIONS, AND EQUIVALENT AOQs		49
APPENDIX C: CSP-A TABLE		51
APPENDIX D: CSP-M TABLES		53
REFERENCES		59
INDEX		69

FOREWORD TO THE SECOND EDITION

The ASQC Basic References in Quality Control: Statistical Techniques has now grown to sixteen volumes, and others are in preparation. It remains a project of the Statistics Division of ASQC. The series' Review Board now consists of Nancy E. Baxter-Belunis, Gerald J. Hahn, Lynne B. Hare, Stuart J. Janis, Norman L. Johnson, Saeed Magh Soodloo, Michael J. Mazu, William M. Mead, Annabeth Propst, Samuel S. Shapiro, Harrison M. Wadsworth, William H. Woodall, and Hassan Zahedi.

The subject matter of the *How to* series continues to be dynamic with contributions by numerous authors and organizations each year. Many of these contributions appear in ASQC journals, hence, periodic revisions of the booklets in the series are encouraged to keep our readers up-to-date on developments. These may be expressed in the booklets by new areas, expanded areas, and/or as additions to the reference sections to permit the interested reader ease of access to the body of knowledge making up each series' subject. This is certainly the case in the revision of Volume 2.

Dr. Kenneth S. Stephens, author of this booklet, has recently retired from the United Nations Industrial Development Organization (UNIDO). His last assignment there was as senior industrial development officer, Institutional Infrastructure Branch, Industrial Institutions and Services Division, Department of Industrial Operations, Vienna, Austria, UNIDO's headquarters. Following retirement from UNIDO he joined the faculty of Southern TECH, Marietta, Georgia, in the Industrial Engineering Technology Department. Earlier in his career, he has been associated with Western Electric, Rutgers University, LeTourneau College, and Georgia Institute of Technology.

Dr. Stephens has worked closely with Harold F. Dodge in the original development of specialty sampling plans. Much of the subject material in this booklet was developed in conjunction with his work with Harold Dodge and Western Electric—hence, he is able to give the reader a clear picture of both the theory and application of these procedures. His expertise in the area, coupled with substantial industrial and consulting experience and academic background, make this booklet one that should be a standard technician's manual as well as a standard classroom and course text and research reference in the area of continuous sampling (CSP) for many years to come.

The second edition includes an updated and comprehensive reference section that should serve the practitioner and researcher alike for ready access to the large literature associated with this subject. In addition to the procedures, schematics, formulas, derivations, tables, nomographs, and examples for numerous continuous sampling plans that were contained in the earlier edition, the revision includes several new developments in CSP procedures and evaluations. A new appendix has been added to resolve a controversy over an apparent anomaly in several of Dodge's earlier papers.

Edward F. Mykytka
Air Force Institute of Technology/ENS
Wright-Patterson Air Force Base, Ohio
March, 1994

FOREWORD TO THE FIRST EDITION

The ASQC Basic References in Quality Control: Statistical Techniques is a literature project of the Statistics Division of ASQC. The series' Review Board now consists of Joseph W. Foster, Norman L. Johnson, H. Alan Lasater, Edward A. Sylvestre, and Harrison M. Wadsworth, Jr., supplemented (for the current volume, second in the series) by Peter R. B. Whittingham.

The current booklet has been field-tested for usability by quality technicians in industry, and was found very suitable.

At the 1979 ASQC Annual Technical Conference (ATC) in Houston, a 1.5 hour tutorial on this booklet (galley proofs of which were distributed to the audience) was presented to over 200 attendees by Dr. Kenneth S. Stephens. In addition Dr. Wayne B. Nelson conducted a 1.5 hour tutorial on his booklet (Volume 1 of this series, *How to Analyze Data with Simple Plots*) for over 200 people.

Dr. Kenneth S. Stephens, author of the current booklet, is a Lecturer in the School of Industrial and Systems Engineering, Georgia Institute of Technology, Atlanta, Georgia. His expertise in the area, coupled with substantial industrial and consulting experience as well as academic background, combine to make this booklet one which should be a standard technician's manual and as well a standard classroom and course text and reference in the area of CSP for many years to come.

Edward J. Dudewicz
Katholieke Universiteit Leuven
Leuven, Belgium
May, 1979

Chapter 1

Introduction

ABSTRACT

Continuous sampling plans were devised for processes involving a continuous or nearly continuous flow of products or other entities. This booklet explains (how to perform) continuous sampling plans. Included are selections of various types of plans, determining parameters, operating the plans, and evaluating the performance of the plans. A brief introduction to acceptance sampling is included. The booklet traces the development of various types of continuous sampling plans. Schematics, nomographs, and tables are included to assist the user. A comprehensive set of references is given to allow for further study and more extensive use of the techniques and principles.

1.0 BRIEF INTRODUCTION TO ACCEPTANCE SAMPLING PLANS AND CONCEPTS

Acceptance sampling is defined as, "sampling inspection in which decisions are made to accept or not accept product or service; also, the methodology that deals with procedures by which decisions to accept or not accept are based on the results of the inspection of samples."* An *acceptance sampling plan* is defined as, "a specific plan that states the sample size or sizes to be used and the associated acceptance and non-acceptance criteria."* Implied in both of these definitions is the notion of *sampling*, in which a quantity of units less than the whole is used to make an acceptance–rejection-type decision affecting the whole. Also implied is the performance of some type of test or evaluation of individual sample units to determine conformity to a set of requirements on these units.

In the following subsections a brief sketch will be drawn of the underlying principles and concepts of acceptance sampling to serve as background information for the main discussion of continuous, skip-lot, and chain sampling. The latter two are the subject of a second ASQC basic reference series booklet by the author, Stephens (1995).

1.1 BASIC ELEMENTS OF A SAMPLING PLAN AND ACCEPTANCE SAMPLING

A principal element is the sample itself, which, according to the sampling procedure, may be of one or more fixed sizes (denoted by n, n_1, n_2, etc.) or may be less explicit and defined in terms of accumulation of individual units to reach a given decision or one or more fraction of units to be selected (denoted by f, f_1, f_2, etc.).

Acceptance implies a decision on a certain quantity of units, which again may be fixed, such as a lot size (denoted by N), or it may relate to individual units, blocks of consecutive units, or, more conceptually, the process producing the units. Acceptance further implies "passing" the associated units or process as opposed to an alternative disposition such as non-acceptance or rejection. That itself may constitute alternatives such as scrapping, sorting, screening or detailing (100% inspection of individual units), repairing, reprocessing, etc.

Acceptance sampling and plans for carrying it out, as defined earlier, combine the notions of a sample with criteria for making decisions. One of the simplest of acceptance sampling plans is the specification of a single sample size n to be drawn randomly from lots of fixed or variable size, and an acceptance number c defined as "the largest number of variants or variant units in the sample that will permit acceptance of the inspected lot on batch."* Due to the wide variety of sampling plans and procedures, for even this simple sampling plan to be described fully, it must be referred to as a "lot-by-lot, attribute, single acceptance sampling plan."

---

*See References, ANSI/ASQC Standard A2-1987 (1987). Additional terms, such as *sample, sample size, unit, inspection,* etc., not explicitly defined in this booklet will follow the definitions and notations of this reference.

The criterion (or criteria) for acceptance is (are) thus another principal element. It also varies considerably according to the sampling procedure, and this is certainly the case for the general type of sampling procedure treated both in this booklet and the one on skip-lot and chain sampling, Stephens (1995).

Following further introduction of the principles of acceptance sampling, a more complete list and discussion of essential items to be included in developing a sampling plan are presented in section 1.5.

1.2 EVALUATION OF THE PERFORMANCE OF SAMPLING PLANS—OC CURVE

Statistical methods, of which acceptance sampling is one, involve controlled risks. Whenever a portion or sample of available units is inspected to determine the acceptability of a larger group of units, risks are taken that the groups will be accepted or not accepted in error. Due to inherent fluctuations in the results of sampled units, groups containing nonconformances at levels considered to be satisfactory may not be accepted. Similarly, groups containing nonconformances at unsatisfactory levels may be accepted. These are sometimes referred to as Type 1 and Type 2 errors, respectively.

What is needed then is an evaluation of these risks for sampling plans under consideration for use. To this end the principles and techniques of mathematical statistics and probability are used. For many sampling plans, particularly of the lot-by-lot type, a common measure of these risks is the probability of acceptance Pa for a given level of nonconformance. By making such determinations over a range of these possible levels, an evaluation of the expected performance of a sampling plan is obtained. When such tabulations of a series of probability of acceptance (Pa) values associated with levels of nonconformance (often expressed in percent or fraction defective* p) are plotted on a graph, with interconnection of the points, the result is known as the Operating Characteristic Curve (OC curve) for the given sampling plan. An example of an OC curve is shown in Figure 1.1 for the attribute single sampling plan with sample size n of 100, and acceptance number c of 4. The probability of acceptance Pa (for this curve) is based on the binomial** probability distribution for different values of process fraction nonconforming p. Such curves can be studied to assess the degree of quality protection afforded by the sampling plans for selection of one having desired characteristics. They are also useful for grouping sampling plans of varying parameters (such as n and c) having common characteristics. For a greater discussion of operating characteristic curves and their determination for common sampling plans, see Duncan (1986) and Grant and Leavenworth (1988).

Some sampling plans and procedures, notably those in this booklet, do not lend themselves totally to evaluation by the ordinary OC curve. In these cases, other measures, often unique to the type of sampling plan, must be devised. In other cases, an OC curve can be determined but may require a modified interpretation, since consecutive decisions may not be independent. Response characteristics—measuring ability of a sampling plan to respond to a change in quality level—provide further evaluation.

1.3 QUALITY LEVEL INDEXING TO CLASSIFY SAMPLING PLANS—AQL, AOQL, LQL

For convenience to users, sampling plans affording certain types of protection are often grouped together and tabulated as such. Tables of sampling plans for different values of the designated protection then provide choices for applications to different products, quality characteristics, types of tests, etc.

Common indices for grouping of sampling plans are Acceptable Quality Level (AQL), Average Outgoing Quality Limit (AOQL), and Limiting Quality Level (LQL)—all of which are defined in ANSI/ASQC Standard A2-1987 (1987). There are others, such as the indifference quality for choice of plan with $p_{.50}$ serving as a point of control (also see the ANSI/ASQC Standard A2).

AQL places emphasis on the producer's risk, i.e., the risk (to the producer) that products of acceptable quality (nonconformances no worse than the AQL) will not be accepted. AQL classification ensures a relatively high probability of acceptance for product of acceptable—or near acceptable—quality.

---

*For definitions of terms such as *defective, defect, nonconformity,* and *nonconforming unit,* see ANSI/ASQC Standard A2-1987 (1987) as well as *Glossary and Tables for Statistical Quality Control,* ASQC (1983). All of these terms are used in various places in this booklet. The reader is advised, in any instance, to substitute whichever term applies to the situations for which the principles and techniques are being used.

**This is exact here since an infinite lot is assumed.

Figure 1.1
OC Curve for n = 100, c = 4

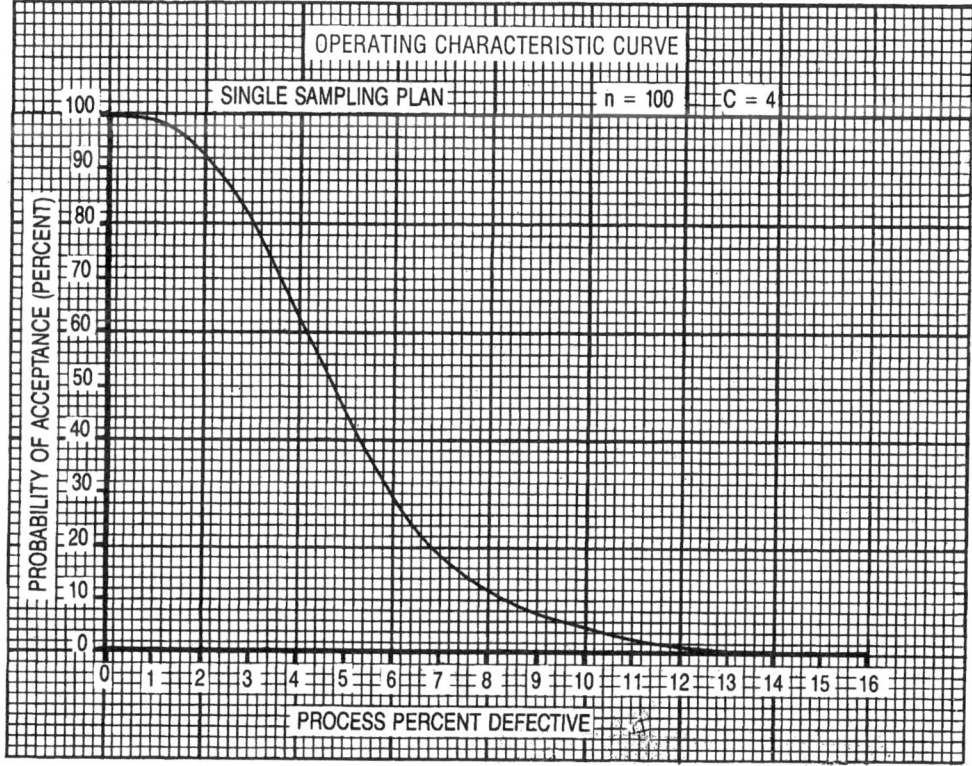

AOQL protection requires a sampling and inspection procedure involving rectification as by 100% inspection of lots not accepted, replacement or correction of variant units found on sample units and sorted units, and a mixing of accepted product with sorted product. Hence it provides average protection over a period of time (lots), but not single lot protection.

LQL places emphasis on the consumer's risk, i.e., the risk (to the consumer) that product of an unacceptable quality (nonconformances as great or greater than the LQL) will be accepted. LQL classification ensures a relatively low probability of acceptance (commonly 0.10) for product of unacceptable quality. LQL has its counterpart in MIL-STD-105E (1989) called *limiting quality* (LQ) with Pa set at either 0.05 or 0.10. Sometimes this point is referred to in the more meaningful term of *rejectable quality level* (RQL).*

1.4 TYPES OF SAMPLING PLANS

The sampling plans in this booklet will be better understood in principle, in application, and in their current state of development when considered among the various types of sampling plans and situations. The purpose of the following subsections is to produce this understanding.

1.4.1 Attributes and Variables Data and Sampling Plans

Two categories of data (observations, measurements) are generally associated with quality control and acceptance sampling methodology. These are attributes and variables data, and are defined in the A2-1987 ANSI/ASQC standard. They relate to counts, proportions, or percentages (attributes) and to measurements

---

*The A2-1987 ANSI/ASQC standard indicates LQL as the preferred symbol, with LTPD and RQL as alternatives.

made on a continuous scale (variables). Other terms used for these quantities are discrete (attributes) and non-discrete (variables). Sampling plans are of the attribute type or the variables type according to the type of data involved. (Mixed attribute–variables sampling plans are also possible and available in the literature.)

1.4.2 Single, Double, Multiple, and Sequential Sampling Plans

Acceptance-rejection decisions can be made on the basis of a single sample by comparing the sample results (attribute or variable) with the acceptance-rejection criteria. However, the criteria associated with sampling plans can also be designed to allow the decision of acceptance or rejection to be postponed until additional samples are inspected. Plans that require a final decision on or before two samples are called *Double Sampling Plans*. Plans permitting more than two samples to reach to a final decision are called *Multiple Sampling Plans*.

Sampling plans may be designed so that a decision to accept, reject, or continue inspection is made after each consecutive unit. Such plans are referred to as sequential or more specifically, *Unit Sequential Sampling Plans*. Commonly, double, multiple, and sequential sampling plans are of the attributes type for which all samples or units are drawn from the same lot under consideration for sentencing by the outcome. Sequential sampling usually entails the smallest average sample size, but this advantage is often offset by administrative difficulties in lot-by-lot sampling.

1.4.3 Lot-by-Lot Sampling, Continuous Sampling, and Cumulative Sampling

The last paragraph in section 1.4.2 refers to lot-by-lot sampling. Units are accumulated into discrete lots, a sample or samples are drawn from a given lot, and the decision based on the sample results is made on the associated lot. Sampling plans that operate on this basis are called *Lot-by-Lot Acceptance Sampling Plans*.

Some production processes do not lend themselves to the formation of discrete lots. Units often flow along a continuous conveyor belt or similar consecutive conveyance, not necessarily moving continuously. Inspection may need to be performed on individual units at discrete locations along this flow. For such situations, even the simple concept of a sample size is not feasible. Plans for this type of sampling problem are referred to as *Continuous Sampling Plans* and are treated in some detail in section 2.0.

A variation in the sampling procedure to carry results over from one sample to another so that a decision on a given lot may be based partially on sample results from two or more lots is known as *Cumulative-Results Sampling*. *Chain Sampling* is such a procedure and is treated in some detail in *Volume 4: How to Perform Skip-Lot and Chain Sampling, Second Edition*, by Stephens (1995).

1.4.4 Systems of Sampling Plans—Tables and Procedures

As indicated in section 1.3, sampling plans are often grouped together in tables for the convenience of users. Examples are MIL-STD-105E (1989) and Dodge and Romig (1959). Additionally, several sampling plans may be used together in a single application with a procedure for specifying conditions under which each is to be used. Such procedures are often referred to as sampling schemes or sampling systems. MIL-STD-105E (1989) and MIL-STD-1235C (1988) are examples. See, also, papers by Stephens (1958) and Stephens and Larson (1967).

Of the sampling plans considered here, numerous tables of plans have been prepared for Continuous Sampling Plans. For Skip-Lot Sampling a standard with tables is now available, ANSI/ASQC S1-1987 (1987). Beyond this single standard for Skip-Lot Sampling, and for Chain Sampling Plans, the inspection planner must design his or her own plans using the information and techniques presented in Stephens (1995) and its references. Standard tables for Chain Sampling have not yet been developed.

1.5 ESSENTIAL ITEMS FOR THE PREPARATION AND DEVELOPMENT OF SAMPLING PLANS

With all of the various types of sampling plans and sampling situations outlined earlier, the task of selecting a sampling plan is not simple. A common error in specifying sampling procedures is to specify (or request specification of) the sample size first. The following sections contain a list of essential items that need to be considered in the preparation and development of sampling procedures for specific cases.

It is significant to note that the actual determination of sample sizes for the sampling procedures is not made until Step 9. The information obtained in the first eight steps is essential to a proper and reasonable determination of sample size.*

In addition to the discussion presented here, including the step-by-step approach to compiling relevant information and making a decision on what sampling plan to use, interested readers should also consult the paper by Godfrey and Mundel (1984). This paper presents some very useful conditions, criteria, process information, consumer needs, etc., while proposing a guide for the selection of an acceptance sampling plan. The technical report (ISO TR 8550:1994) alluded to in that paper should also be consulted.

1.5.1 Preliminary Quality and Test Information

1. Determine and list each *quality characteristic* by name and description. This can be done by considering all possible defects that may affect the product or its usage.
2. Determine and specify the *product unit* to which each quality characteristic applies and upon which the test (of Step 3 following) is made. Different product units may apply to different quality characteristics on the same product. Each needs to be specified clearly.
3. Develop and specify the *test method* to be used on each product unit for testing the product for conformance to each quality characteristic.
4. Determine and specify the *criteria for conformity* for each quality characteristic under test. Give a clear, detailed description of when a single product unit has or has not met the specification of the quality characteristic. For example, what criteria must a single product unit satisfy in order to be acceptable for each quality characteristic under test, or, conversely, what constitutes a defect or departure from the requirements of the quality characteristic? Specification limits or other requirements for quality characteristics must be compatible with the sample criteria referenced in Step 9(e).

1.5.2 Refined Quality and Acceptability Information

5. Determine and list the classification of defects and/or groups of defects that affect a unit of product. Classification will refer to the degree of importance or severity of the defects, such as minor, major, or critical. Further classification, isolation, or grouping of defects (with the severity classes) shall be made with respect to the cost and/or complexity of testing (as per Step 3) including extensive environmental tests, life test, destructive tests, etc., for special handling in sampling inspection. Different classifications or groupings may be necessary for different product units as in Step 2.
6. Establish an AQL or other appropriate quality level (AOQL, LQL, etc.) upon which to base the sampling procedures for each defect classification, defect grouping, or combination (as per Step 5). The purpose of Step 5 is to establish a relative importance between the product defects, while the purpose of Step 6 is to assign somewhat of an absolute measure of quality to the product through its defect classifications and groupings.
7. Determine if sampling and acceptability should be based on lot inspection, on-line (continuous) inspection, in-process inspection, bulk inspection, etc., for each defect classification, defect grouping (as per Step 5), or product unit distinction (as per Step 2). Obtain appropriate information such as composition of lots, common lot sizes and ranges of lot sizes, production rates per time unit for continuous processes, sublot packaging units, etc.

1.5.3 Determining Sampling Methods and Sampling Plans

8. Combining and using all of the information in Steps 1–7, decide what method and type of sampling shall be used for each defect classification, defect grouping, and/or product unit distinction. The nature of the production process needs to be considered in these decisions. Steady, high-volume production may permit application of sampling procedures not valid for job–shop-type production of isolated, one-off lots. The nature and extent of the process control utilized in the process will

---

*Portions of the following are contained in Stephens (1976) and are used here by permission of APO.

also affect the sampling procedures chosen. The degree of accuracy desired and the workload required by the sampling and testing also need to be considered.

9. Develop the sampling plans, including
   (a) Composition and sizes of the inspection lots, batches, bulk product, etc.
   (b) Sample sizes, including, if applicable, multistage subsampling as to how many sublots to sample in a lot, how many subsamples to include from each sublot, how many units to include in each subsample, how many test specimens to prepare from each unit, etc.
   (c) If applicable, the relationships between lot size and sample size and the preparation or specification of an appropriate sampling table.
   (d) Whether to test each product unit individually, in groups, or as a composite.
   (e) The criteria for conformity of the sample—acceptance numbers for defectives, defects, etc.; critical values for averages, ranges, standard deviations, etc.
   (f) Special sampling system procedures such as tightened, normal and reduced inspection plans, skip-lot inspection, multi-level plans, etc., with the objective of saving work and providing better quality protection and discrimination for the consumer and producer.
   (g) Disposition of rejected lots or product groupings, provision for resubmission of reworked product, etc.

1.6 ARE SAMPLING PLANS PASSÉ?

For the current revision of this volume and at this juncture in the development and evolution of the quality sciences, it is appropriate to raise the question and provide some answers as to whether sampling plans have become passé. A certain class of authors and proponents of some approaches would have you believe so. Fortunately these views are also offset by others that understand the totality of the quality sciences and the importance of utilizing (or at least having available) a broad range of tools and methodologies as operational needs require. The quality sciences have always been plagued by problems of semantics. Young (and/or new) proponents of total quality management (TQM) often are ignorant of, or ignore, the fact that programs and systems with previous names as simple as quality control or total quality control (and existent as much as 25–35 years ago) included such concepts and methodologies as project-by-project continuous improvement *with a prevention orientation and with quality teams,* customer needs assessments and satisfaction programs, quality as a strategic business component including its contribution to costs and cost reduction, design quality and innovation, etc. This is not to say that important strides in refining and exposing these concepts to a wider audience have not been made in recent years; they have. But overzealous proponents of certain concepts have shown tendencies to idealize and ignore the conditions and necessities calling for a full range of statistical (and other) tools for the *total* job of achieving quality and its related benefits.

While in the ideal world we should produce a product or offer a service that emanates from a process that is in statistical control at an acceptable level, the truth is that in many instances this is not the case. Thus, in reality, sampling, and more specifically, acceptance sampling plans, are still useful in many applications—most often involving assurances of process quality levels that cannot be assumed to be in a state of statistical control and consistent or subject to latent and/or environmental (or even fraudulent) change. Certainly, they are not a panacea, nor does this volume or many like it propose such. They are among the tools that discerning, knowledgeable practitioners will use along with other tools directed more at planning, prevention, and improvement. Hence, certain prevalent negative attitudes toward acceptance sampling are much exaggerated and appropriate only in a perfect environment.

Harold Dodge, as founding father of acceptance sampling, was always a proponent of the full and proper use of acceptance sampling plans that involves the feedback of results for action on the process or product improvement. See, for example, Dodge (1948). Dodge was also a member of the Bell Laboratory team that devised and promoted many of the original quality control tools, including the control chart. He was one of the early users of control charts to control and improve processes and provide feedback results from inspection applications. See Wadsworth, Stephens, and Godfrey (1986), page 115.

Recent viewpoints on retaining the full tool kit (that includes acceptance sampling plans) for modern quality programs are expressed in Schilling (1990–91), Schilling (1991), and Milligan (1991). See also the 1990–91 issue of *Quality Engineering*, Vol. 3, No. 2, pp. vii–xii. The arguments presented are relevant and persuasive and need not be repeated here. Further relevant discussion of this topic is contained in the papers (and their numerous references) by Sower, Motwani, and Savoie (1993), and Taylor (1994).

Chapter 2

Continuous Sampling

2.0 BRIEF INTRODUCTION TO CONTINUOUS SAMPLING PLANS (CSP)

As described in section 1.4.3, sampling procedures have been devised for handling situations involving a continuous flow of product and are known as Continuous Sampling Plans (CSP). These plans are used for production processes where no separate lots are formed. They are generally used on some type of conveyor but are applicable to any continuous type operation where it is not desirable to accumulate the product into lots for purposes of inspection. Continuous Sampling Plans were first developed by Dodge (1943), (1947) and subsequent developments represent extensions and variations of his basic procedure.

2.1 BASIC PARAMETERS, OPERATION, EVALUATION, AND APPLICATION (CSP-1)

Continuous sampling plans are sometimes referred to as random order plans because the theory on which they are based has to do with the occurrence of defects in a random series of units. The theory in most respects is similar to that used in lot-by-lot inspection, although the terminology and some of the inspection practices are different. Three common principles upon which the development of sampling inspection procedures, including continuous sampling plans, depends are as follows:

1. When samples or individual units are taken from product of a certain quality (as measured in fraction or percent defective or defects per unit), the occurrence of defects forms a certain statistical pattern.
2. When samples or individual units are taken from product of a different quality, the occurrence of defects will form a different statistical pattern.
3. Because of this difference in pattern, it is possible to set up acceptance criteria for sampling inspection that will reject more product (even to a predictable extent) when the quality is worse and accept more product when the quality is better.

Suppose we have a continuous flow of product that is 4% defective. We begin to inspect this product classifying each unit in order as defective or non-defective. If O represents a non-defective unit and X a defective one, the record of consecutive inspection results might be similar to the following:

XOOOOOOOOOOOOOOOOOXOOXOOOOOOOXOOOOOOOXOOOOOOOOOOOOOX . . . .

The number of units from the occurrence of one defective to the occurrence of the next defective is a statistic following a probabilistically predictable pattern and may be referred to as the defective spacing s. In this example, s is equal to 18, 3, 8, 9, 14, etc. These values of s can be plotted like any other series of numbers from a process and will form a fluctuating statistical pattern. If product is worse than 4% defective, the defective units will occur more frequently and s will tend to become shorter. If product is less than 4% defective, the defective units occur less often and s tends to become longer. Figures 2.1, 2.2, and 2.3 are records each of some 15 observed values of s from product that was 2%, 4%, and 8% defective, respectively.

Since product of different quality produces different patterns of s, it is possible to set up an acceptance criterion in terms of s that will reject more product of a bad level of quality and accept more product of a good level of quality—where *good* and *bad* may be defined as desired and varied for different applications.

Parameters. How this is actually done for continuous sampling plans varies with several types of plans. The first plan by Dodge (1943), (1947), later designated CSP-1 by Dodge and Torrey (1951b), specifies a clearing interval i, which is a fixed parameter for a given continuous sampling plan, denoting the number of consecutive units to be inspected and found clear of defects before the process qualifies for sampling.

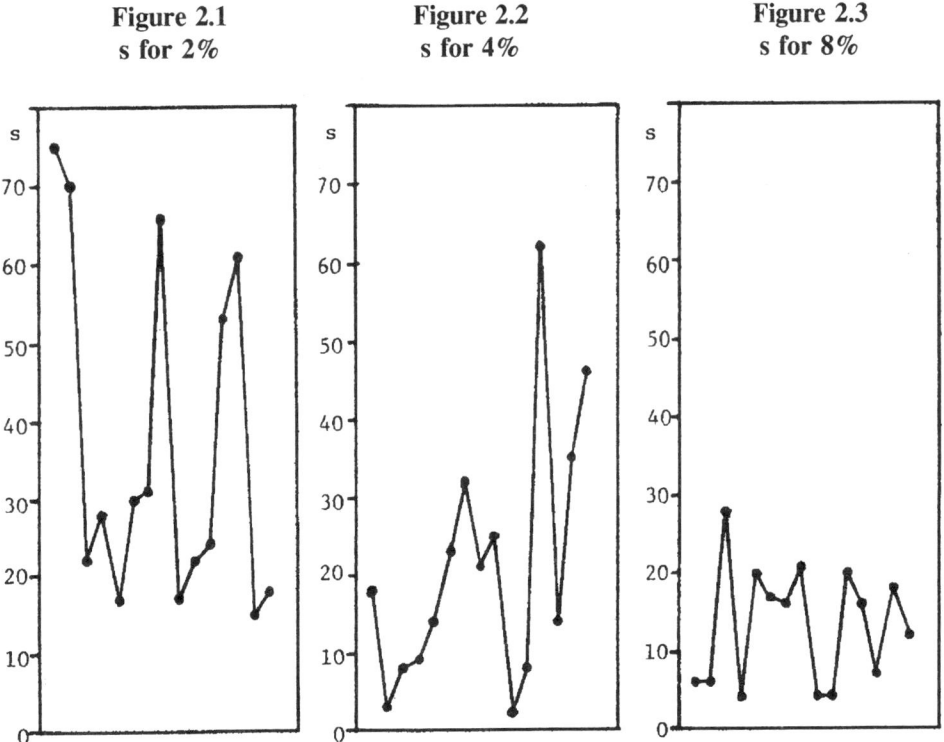

Figure 2.1
s for 2%

Figure 2.2
s for 4%

Figure 2.3
s for 8%

Initially, consecutive units are inspected 100% until the clearing interval i qualification is met. After such qualification, only a fraction f of the units is inspected, selecting individual units one at a time from the flow of product, in such a manner as to ensure an unbiased sample of fraction f (this will be considered in greater detail later).

When a unit is found defective (application is also possible to individual defects or classes of defects as detailed later) during the sampling period, immediate reversion to 100% inspection is again required until qualification for sampling by again satisfying the clearing interval i. Continuous sampling is generally of the AOQL type (see section 1.3 and section 2.2.2) involving periods of 100% inspection and periods of sampling. The AOQL achieved is determined by the values of i and f chosen.

In this example we might decide to use a clearing interval of 20. If the product is 2% defective, about 67% of the defective spacings (measured by $s - 1$, the length of a series of non-defective units) would meet this criterion allowing sampling of f fraction of units quite often. If the product is 4% defective, only about 44% of the spacings would meet it, and for 8% defective, only 19% would meet it. Thus, we have an acceptance criterion that is able to distinguish between product of different quality and does not depend on forming the product into separate lots. In Figure 2 of his original paper, Dodge (1943) presents curves defining the probability distribution of random order spacing of defects in uniform product.

There are two other ways in which continuous sampling differs from lot-by-lot sampling.
1. In continuous sampling there is no fixed sample size. The sample is expressed as a fraction or percentage of the continuous flow of product. For example, the sample may be expressed as 10% of the product, denoting an f of 0.10 of 1/10. This can mean different things. At least three ways of applying the parameter f to the drawing of sample units have been identified. These are (a) *systematic sampling:* inspect every 1/f th unit produced; (b) *block* or *random sampling:* inspect a randomly chosen unit from each successive block, lot, or segment of 1/f units produced; and (c) *probability sampling:* inspect randomly selected units so that f is the overall average frequency of inspected units. Any of these methods of sampling may be used, taking into account the nature

**Figure 2.4
Operation Schematic for CSP-1**

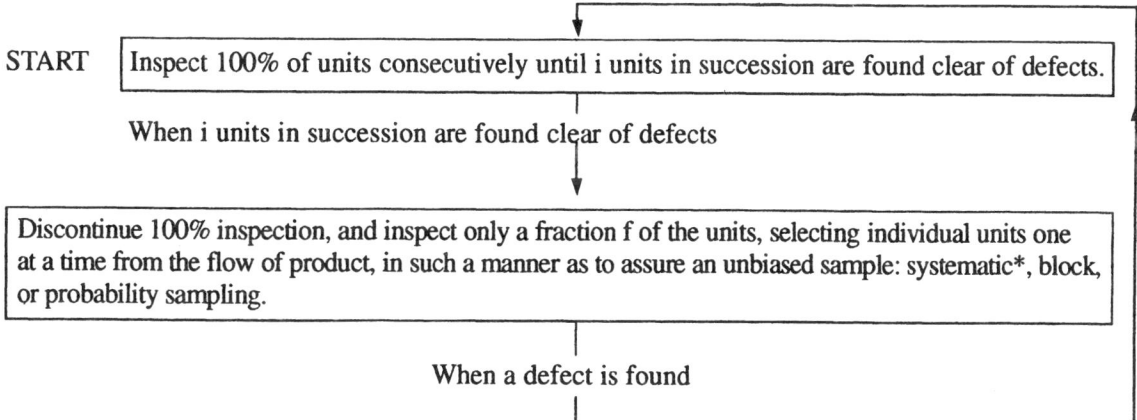

*When consecutive units in the process are *not* prearranged so that the sample unit has been preinspected or given extra care in its fabrication.

of the process to which applied. See Dodge and Torrey (1951b); Murphy (1959a) and (1958); and Derman, Johns, and Lieberman (1959) for additional discussion on this matter.

2. The OC curve (see section 1.2) for a continuous sampling plan is not the same as for a Lot-by-Lot Sampling Plan. In lot-by-lot sampling inspection (of the AOQL type) the probability of acceptance refers to (1) the probability of accepting a single lot without having to inspect 100% or (2) the proportion of lots that will be accepted out of a stream of lots. In continuous sampling there are no specific lots. Consequently a different measure of evaluation of the performance of CSP is needed. The most common measure is the percent of total production accepted on a sampling basis as a function of the incoming percent defective. Other measures of the performance of CSP-type plans are (a) average number of units inspected in a 100% screening sequence, (b) average number of units passed under the sampling procedure, (c) average fraction of total produced units inspected in the long run, and (d) average outgoing quality. Still further measures of evaluation include spotty quality: Dodge and Torrey (1951b); stopping rules: Murphy (1958) and (1959a), Magwire (1956), and LeMaster and McKeague (1958); *minimum* average fraction inspected: Resnikoff (1960); lack of control: Derman, Johns, and Lieberman (1959), and Hillier (1964a); finite length production runs: Brugger (1972a), Blackwell (1977), McShane (1989), and McShane and Turnbull (1991); robustness: Lasater (1970); unrestricted AOQLs: Lieberman (1953), White (1964) and (1965), Endres (1967c) and (1969), Brugger (1967) and (1976), Banzhaf and Brugger (1970), and Sackrowitz (1975); inspection error: Case, Bennett, and Schmidt (1973); non-Bernoulli, Markov incoming quality models: Kumar and Rajarshi (1987), McShane (1989), and McShane and Turnbull (1991); probability limits on AOQL and estimates of incoming quality: McShane and Turnbull (1991); and economic design: Chui and Wetherill (1973), Gregory (1956), Satterthwaite (1949), and Savage (1959). Many of these papers address one or more of the additional measures of evaluation, as shown for some cases above. The interested reader should consult the literature for a study of these additional evaluations of CSP in order to have further insight into and understanding of the procedures. They range from pathological, extreme scenario situations to very practical and informative conditions that shed light on the operation of CSP schemes.

Operation. A simple operation schematic of CSP-1 is shown in Figure 2.4. Not explicit in this and subsequent schematics is the option to correct or replace defective units with good units *or* to simply remove defective units (nonreplacement). Operation schematics for various extensions of CSP-1 are given in section 2.2.

Selection of the parameters for CSP-1, i.e., AOQL, i, and f, is possible in several ways. This is given in section 2.3 (via nomographs) and section 2.4 (via tables) for both CSP-1 and its extensions.

Evaluation. As indicated earlier, evaluation of CSP procedures differs from lot-by-lot procedures. Specific plans can be evaluated via direct calculations using formulas developed by referenced authors, some of which are given in this guideline. Modern handheld scientific calculators are ideally suited to these calculations. Some specialized and some general computer programs have been developed for the basic procedures and characteristics. Early developments involved programs for the Monroe desktop and the Texas Instruments, TI-59 handheld, programmable calculators. A FORTRAN program is given by Sheesley (1975). Banks (1989) provides a BASIC program for IBM personal computers (and compatibles) on pages 541–544 with examples of outputs on pages 538–540. A FORTRAN program is given by McShane and Turnbull (1992) that "computes new performance measures for continuous sampling plans applied to finite production runs."

For CSP-1, the five most common measures of performance are

1. The average number of units inspected in a 100% screening sequence following the finding of a defective unit u, namely,

$$u = \frac{1 - q^i}{pq^i} \qquad (2\text{-}1)$$

where p is the process fraction defective and q = 1 – p.

2. The average number of units passed under the sampling procedure before a defective unit is found v, namely,

$$v = \frac{1}{fp} \qquad (2\text{-}2)$$

3. The average fraction of total produced units inspected in the long run, F,* namely,

$$F = \frac{u + fv}{u + v} = \frac{f}{f + (1 - f)q^i} \qquad (2\text{-}3)$$

4. The average outgoing quality, AOQ, namely,

$$AOQ = p(1 - F) = \frac{p(1 - f)q^i}{f + (1 - f)q^i} \qquad (2\text{-}4)$$

5. The average fraction (or percent) of total production accepted (passed) on a sampling basis P_a, namely,

$$P_a = \frac{v}{u + v} = \frac{AOQ}{p(1 - f)} = \frac{1 - F}{1 - f} = \frac{q^i}{f + (1 - f)q^i} \qquad (2\text{-}5)$$

These formulas apply when defective units found are corrected or replaced with good units (known as the replacement case). Dodge (1943) has noted that the substitution of i by (i – 1) yields the formulas for the nonreplacement assumption. When referring to specific CSP-1 plans with parameters i and f and related AOQ characteristics, Dodge (1955) further mentions, "It can be shown that i should be increased by one in CSP-1 plans when defective units are removed but not replaced." His Procedure A1 (the replacement case) uses i, whereas his Procedure A2 (the nonreplacement case) uses (i + 1). Since these developments and related statements by Dodge in the cited papers (of 1943 and 1955), there have been occasions when these statements were thought to be contradictory and in some cases incorrectly applied. This subject is considered in greater detail in Appendix B where it is shown that both statements are correct, and the applicable conditions for each case are clearly elucidated.

---

*Referred to as average fraction inspected (AFI) in some literature. See, for example, section 2.4.1.

The fifth performance measure is often referred to as the operating characteristic (OC) function. (Strictly speaking, all of them are operating characteristics.) It is used as the OC curve for CSP procedures (see section 1.2).

Other evaluations are discussed as related to the topics of sections 2.2, 2.3, and 2.4. The reader interested in more comprehensive knowledge of CSP evaluations is referred to the literature given in the References.

Application. CSP procedures are designed for application to situations "where it is neither convenient nor practical to group product articles in collective lots or batches for the purposes of inspection," Dodge (1970). They are generally applicable to in-line and end-of-line inspections and have been found "to be most effective when administered in such a way as to provide an incentive to clear up the faults in a process promptly." This is commonly achieved by requiring the production organization to perform the 100% inspection via a screening crew when the results dictate. An acceptance inspector performs the sampling inspection and commonly continues to sample even during periods of 100% inspection by the screening crew as a follow-up on its effectiveness (there being many demonstrations and papers on the ineffectiveness of some 100% inspections). This procedure has been incorporated in systems of continuous sampling plans as illustrated in section 2.4.

Dodge (1943), (1970) conceives of two procedures referred to as A and B, with A applicable to a product of consecutive articles and B to material offered as a flow of consecutive lots or sublots of articles. Both procedures, under CSP-1, follow the simple diagram of Figure 2.4. Necessary conventions are to treat the order of production of procedure A with order of inspection in procedure B, since strict order of production by articles is lost in sublotting. Additionally, in procedure B, when it is necessary to find i inspected units in succession clear of defects, the 100% inspection must be allowed to extend to the immediately succeeding lots, if i units in succession are not found in the current lot.

Other considerations in application of CSP procedures are

1. Ample space, equipment, and personnel at or near the site of inspection to permit rapid 100% inspection when required. Perhaps, as indicated, this will be performed by screening crews who will oscillate between production work and inspection work as the sampling inspection results dictate.
2. Relatively easy and quick inspection, for example, attribute inspection and automatic testing.
3. A process that is producing, or is capable of producing, material whose quality is stable.
4. The inspection is nondestructive, since the procedure incorporates 100% screening.
5. CSP can apply to individual defects, classes of defects, or defective units. When applied to classes of defects, it is possible for one class of defects to be undergoing a clearing interval for the defect that caused rejection, while the other class (classes) of defects would still be under sampling. An additional convention is that if, during a clearing interval, a defect is found that belongs to the class still undergoing sampling, then the discovery of the defect (and its removal) should not be used to trigger a shift to 100% inspection for the class of defects still undergoing sampling. (For example, note the schematic for CSP-A in Figure 2.7, though this feature is applicable to all CSP procedures.)

Application can be to various entities, namely end items, components, raw materials, data or records, persons, and even lots.*

2.2 ADDITIONAL CONTINUOUS SAMPLING PLANS

In addition to the basic procedure referred to as CSP-1 in section 2.1, a considerable number of extensions and variations of continuous sampling plan procedures have been developed. Some will be discussed here and in sections 2.3 and 2.4. Consult the Reference section for a more comprehensive study. Fairly good chronologies of continuous sampling plan development are given by Dodge (1970) and Banzhaf and Brugger (1970) in the same issue of the *Journal of Quality Technology.*

---

*Skip-lot sampling is based on the application of continuous sampling procedures to lots as the inspected entity. See Stephens (1995).

Figure 2.5
Operation Schematic for CSP-2

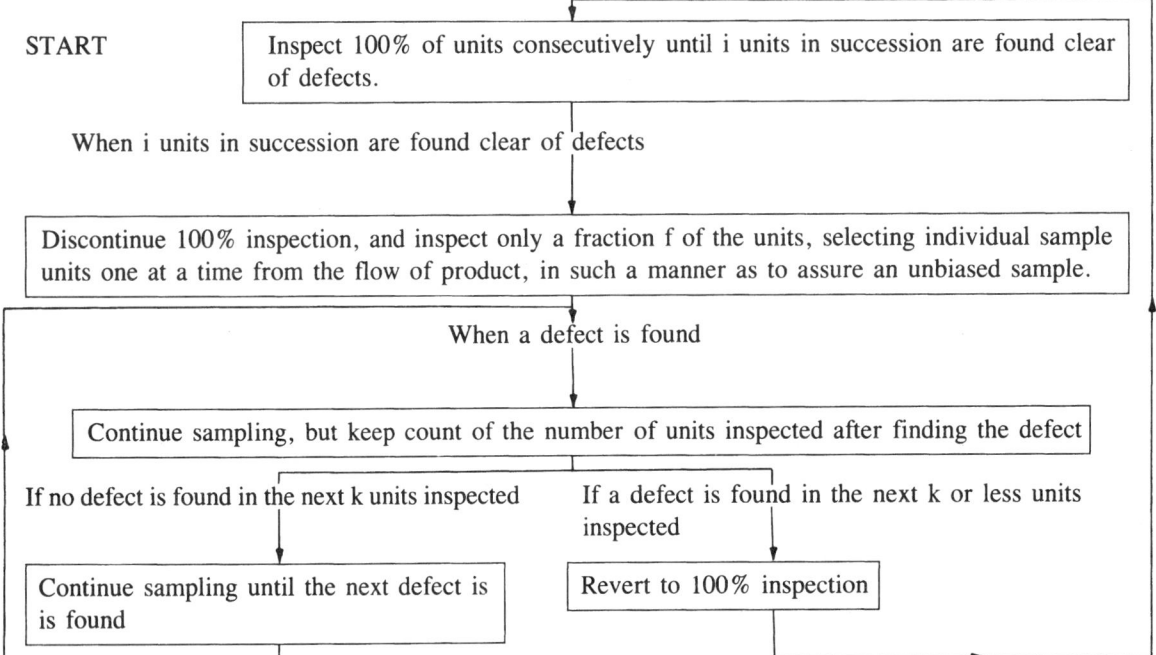

2.2.1 CSP-2 and CSP-3

Extensions of CSP-1, which have received considerable attention and use, are those by Dodge and Torrey (1951b) labeled CSP-2 and CSP-3. As indicated in their paper, both plans grew out of suggestions by inspection personnel engaged in applying continuous sampling.

CSP-2 is to continuous sampling as an acceptance number c greater than zero is to lot-by-lot sampling. That is, it allows for sampling to continue with the occurrence of an occasional defect (defective unit), provided that a defect (defective unit) does not occur too frequently. For given AOQL and f, an increase in i is necessary to compensate for the allowance of a defect. An additional parameter, k, is introduced as the minimum number of consecutive sampled units that must be free of defects, after the occurrence of a defect, for sampling to continue. That is, the 100% inspection phase is reinstituted only if defects occur on sampling with spacing less than k.

An operation schematic for CSP-2 is shown in Figure 2.5, similar to that of Figure 2.4 for CSP-1. This is due to Dodge and Torrey (1951b) and Dodge (1970).

CSP-3 introduces a simple and effective refinement of CSP-2 designed to provide extra protection against the case of spotty quality, i.e., the clustering of excessive defectiveness. Following the occurrence of a defect (defective unit) during sampling, the next four consecutive units are inspected and required to be free of defects for sampling to continue on the CSP-2 basis. Otherwise the 100% inspection phase is invoked immediately.

An operation schematic for CSP-3, again by Dodge and Torrey (1951b) and Dodge (1970), is shown in Figure 2.6.

Evaluation of CSP-2 and CSP-3 is similar to that for CSP-1. Only slight variations in some symbols and definitions are required. The others are the same as for CSP-1.

Figure 2.6
Operation Schematic for CSP-3

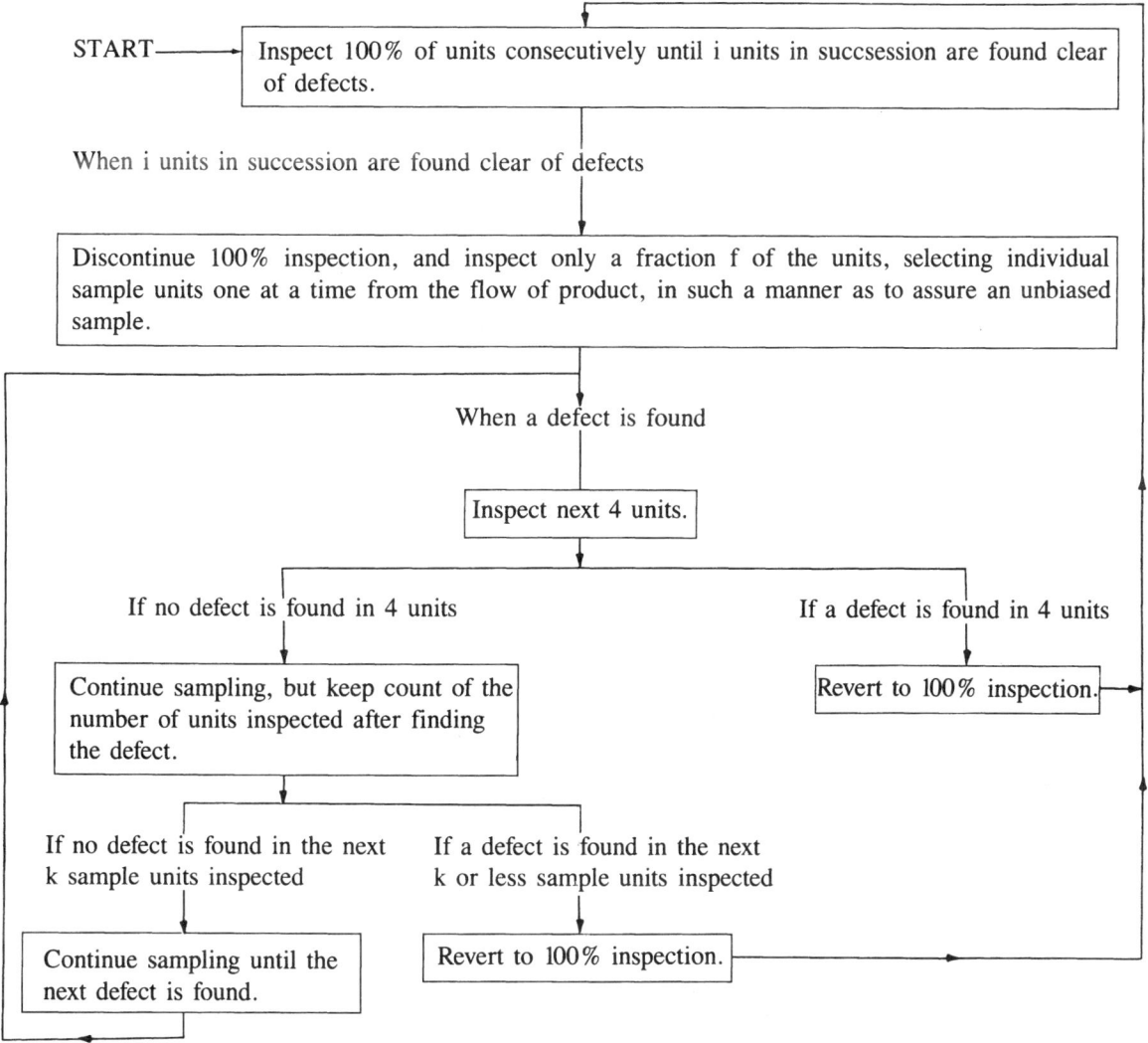

For CSP-2

1. Here u is defined as the average number of units inspected on a 100% inspection basis, namely,

$$u = \frac{1 - q^i}{pq^i} \text{ (same as for CSP}-1). \tag{2-6}$$

2. Here the definition of v is modified slightly to account for allowing a defective unit during sampling and is simply, the average number of units passed during sampling inspection, namely,

$$v = \frac{2 - q^k}{(1 - q^k)fp} \tag{2-7}$$

3. As defined for CSP-1,

$$F = \frac{u + fv}{u + v} = \frac{f(1 - q^k)(1 - q^i) + fq^i(2 - q^k)}{f(1 - q^k)(1 - q^i) + q^i(2 - q^k)} \tag{2-8}$$

4. As defined for CSP-1,

$$AOQ = p(1 - F) = \frac{p(1 - f) q^i (2 - q^k)}{f(1 - q^k)(1 - q^i) + q^i(2 - q^k)} \tag{2-9}$$

and when $k = i$, as commonly practiced,

$$AOQ = \frac{p(1 - f) q^i (2 - q^i)}{f + (1 - f) q^i (2 - q^i)} \tag{2-10}$$

5. Also as defined for CSP-1,

$$P_a = \frac{v}{u + v} = \frac{AOQ}{p(1 - f)} = \frac{1 - F}{1 - f} \tag{2-11}$$

$$P_a = \frac{q^i(2 - q^i)}{f(1 - q^k)(1 - q^i) + q^i(2 - q^k)} \tag{2-12}$$

and when $k = i$,

$$P_a = \frac{q^i(2 - q^k)}{f + (1 - f) q^i (2 - q^i)} \tag{2-13}$$

For CSP-3
1. Since an additional four consecutive units are inspected upon finding a defective on sampling (often referred to as the rule of four), this can be incorporated into u, and it is advantageous to do so (for consistency of results). Hence, let u be the average number of units inspected on a 100% inspection basis including the rule of four, namely,

$$u = \frac{f(1 - q^{k+4})(1 - q^i) + 4fpq^i}{(1 - q^{k+4})fpq^i} \tag{2-14}$$

2. As defined for CSP-2,

$$v = \frac{1 + q^4(1 - q^k)}{(1 - q^{k+4})fp} \tag{2-15}$$

3. As defined for CSP-1 and CSP-2,

$$F = \frac{u + fv}{u + v} = \frac{f(1 - q^{k+4})(1 - q^i) + fq^i(1 + q^4(1 - q^k)) + 4fpq^i}{f(1 - q^{k+4})(1 - q^i) + q^i(1 + q^4(1 - q^k)) + 4fpq^i} \tag{2-16}$$

4. As defined for CSP-1 and CSP-2, $AOQ = p(1 - F)$

$$AOQ = \frac{p(1 - f) q^i (1 + q^4(1 - q^k))}{f(1 - q^{k+4})(1 - q^i) + q^i(1 + q^4(1 - q^k)) + 4fpq^i} \tag{2-17}$$

5. Also as defined for CSP-1 and CSP-2,

$$Pa = \frac{v}{u + v} = \frac{AOQ}{p(1 - f)} = \frac{1 - F}{1 - f} \tag{2-18}$$

$$Pa = \frac{q^i(1 + q^4(1 - q))}{f(1 - q^{k+4})(1 - q^i) + q^i(1 + q^4(1 - q^k)) + 4fpq^i} \tag{2-19}$$

Most* of these results agree with Dodge and Torrey (1951b). Derivation is by means of the steady state or equilibrium probabilities of the states of Markov chains associated with CSP-1, CSP-2 and CSP-3,

---
*The AOQ and Pa results for CSP-3 reflect corrections, with derivation shown in Appendix A.

as shown in Appendix A. Again, correction or replacement of defective units is assumed, as is probability sampling.

2.2.2 CSP-A

With roots in NAVORD OSTD-81 (1952), this sampling procedure was the first to incorporate a stopping rule on the basic procedure of CSP-1. It was carried to Handbook H-107 (1959) and later to MIL-STD-1235 (ORD) (dated 17 July 1962) and although dropped in the subsequent revisions MIL-STD-1235A (dated 28 June 1974), MIL-STD-1235B (dated 10 December 1981) and MIL-STD-1235C (1988), it is included here as an alternative selection for continuous sampling plan applications.

While based on CSP-1, it differs in at least two significant ways, namely,

1. It is applied to a fixed period of production, with the 100% inspection (screening) phase carried out at the beginning of each production period or interval regardless of the state of sampling or screening at the end of the previous period. The production interval is generally associated with a period of time, such as a shift or a day, though estimates of the number of units or product produced in a production interval are needed for selecting the parameters of a specific plan.
2. It incorporates a stopping rule based on limiting the total number of defectives found during screening and sampling within the production interval. A parameter, a, denotes this maximum allowable number of defective units of product (for the designated defect or defects). Hence, from paragraph 8.5.4. of MIL-STD-1235 (ORD), "Whenever, during a production interval, the screening crew and the sampling inspector find a + 1 defective units, inspection will be suspended until the cause of the high rate of defect(ive)s is located and corrected by the supplier, all units already produced but not inspected shall be screened and presented as resubmitted material, and a sampling plan reflecting the new production rate will be used during the remainder of the production interval. The consumer may, at his discretion, accept screening inspection by the supplier as adequate corrective action where the cause of the high defect(ive) rate is difficult and time-consuming to locate and correct."

Other features of the tabled CSP procedures such as CSP-A, which elevate the procedures from sampling plans to sampling schemes or systems, (see section 1.4.4) are (1) use of verification sampling even during the screening phases as a check on the effectiveness of the screening crew, (2) provision for use of a tighter sampling plan upon evidence of poorer quality, (3) provision for use of a reduced sampling frequency upon evidence of good quality, stable and homogeneous production, etc. Details on these topics are given in section 2.4.

Magwire (1956) and Okano and Wolman (1956) report on the determination of parameters and evaluation of sampling plans of this type. Parameters tabled by AQL* (referenced to the AOQL of the plans) and OC curves are given in MIL-STD-1235 (ORD). Use of the tables for CSP-A is discussed in section 2.4.3.

An operation schematic for CSP-A is given in Figure 2.7, based on MIL-STD-1235 (ORD). A table of CSP-A plans from MIL-STD-1235 (ORD) is given in Appendix C, since this standard has been revised (as mentioned). Examples are developed in section 2.4.3 using the tables.

2.2.3 CSP-M

Multilevel continuous sampling plans stretch back to the development work of Lieberman and Solomon (1955) generating a whole series of papers and reports on the new procedures, some of which are Lieberman (1955), Gessford (1955), Resnikoff (1956), Ireson (1956), Bowker (1956), Derman, Littauer and Solomon (1957), Guthrie and Johns (1958), Lieberman and Bowker (1958), Ireson and Biedenbender (1958), Dodge (1960), and Elfving (1962). They were quickly incorporated into manuals of sampling tables and procedures by the Air Force, AMC Manual No. 74-23 (1956); Department of Defense, H-106 (1958); and Department of the Army, MIL-STD-1235 (ORD), dated 17 July 1962, and QSTAG 340 (1974).

---

*For a cogent discussion on the use of AQL values for CSP procedures, see Duncan (1986), pp. 407–413.

Figure 2.7
Operation Schematic for CSP-A

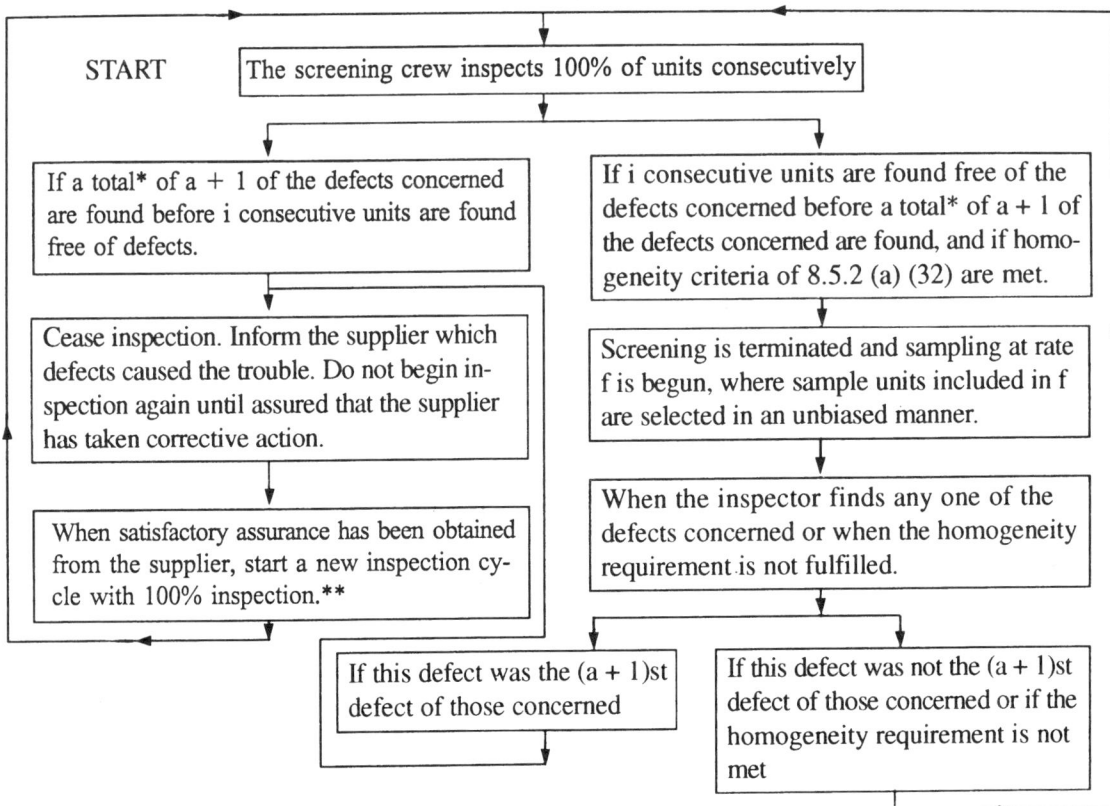

*The total of a + 1 defects includes results of all inspection for the production interval, both sampling and screening.
**For critical defects, the screening should begin with the unit of product just after the last defect-free sample unit.

Other developments were by White (1966), Hassan (1969a, 1969b), Banzhaf and Brugger (1970), Fordice (1972), Sackrowitz (1972), and Kumar (1984).

The designation CSP-M is from MIL-STD-1235 (ORD), but the plan is carried over from Handbook H-106 (1958). Like CSP-A it was dropped in the subsequent revisions of MIL-STD-1235, but is included here again as an alternative selection for continuous sampling plan applications.

CSP-M differs from the original multilevel continuous sampling plans principally in the incorporation of the rule of four from CSP-3 as a protection against spotty quality that may otherwise be difficult to detect as higher sampling levels (being powers of the original fraction f) result in infrequent or reduced sampling. Its basic parameters are i, f, and K, the latter being the maximum number of levels of reduction in the sampling fraction, and, specifically, the maximum power of f employed as a sampling fraction. Based on good quality performance, as measured by the sampling procedure itself, reductions in sampling from f to $f^2, \ldots, f^k, \ldots, f^K$ are permitted, where k = 1, 2, ..., K, and with K varying between 1 and 5 in increments of 1 for specific plans. For K = 1, CSP-M becomes CSP-3.

CSP-M requires that the inspection rate be increased (lower power of f) whenever a defective unit is found during sampling inspection *and* either one of the next four units to pass the inspection station is defective *or* one of the next i – 4 units inspected is defective. The increase in inspection rate is not required if a defective unit is found during sampling inspection *provided* that all of the next i – 4 units sampled and inspected (including the 4 consecutive units to follow the defective unit to the inspection station) are found to be defect-free. On the other hand, a reduction in inspection rate (higher power of f to maximum of K) is permitted if the first i units sampled and inspected at the frequency f^k are defect-free (k = 1, 2, ..., K – 1) or if a defective unit is found while sampling at the frequency f^k and the next 4 consecutive units are defect-free and the next i – 4 sampled (at f^k) and inspected units are defect-free (k = 1, 2, ..., K – 1).

Use of the tables of parameters for CSP-M by AQL (referenced to the AOQL of the plans) is discussed and demonstrated in section 2.4.4. AOQ curves are given in MIL-STD-123 (ORD). The tables also include, other than the built-in tightened and reduced sampling, features such as (1) verification sampling and (2) stopping rule during the 100% inspection phase. Details are given in section 2.4.4.

An operation schematic for CSP-M is shown in Figure 2.8, based on MIL-STD-1235 (ORD). Tables VIII, IX, XI, and XII from the original standard are given in Appendix D. Two examples are developed in section 2.4.4.

2.2.4 CSP-T

The multilevel CSP of MIL-STD-1235C (1988) is designated CSP-T and follows earlier development work on tightened multilevel plans, see for example, Derman, Littauer, and Solomon (1957). The principal difference is a return to 100% inspection upon finding a defective unit at any of the sampling levels. It is further simplified over CSP-M by having the number of sampling levels fixed at three. Additionally, the sampling rates are reduced geometrically by 1/2 between the levels rather than exponentially as in CSP-M. Hence, sampling fractions of f, f/2 and f/4 are used. MIL-STD-1235C gives tables of i and f for AQL indices and designated AOQLs (Table V-A, page 104-4). In addition, CSP-T employs a stopping or warning rule on the maximum number of units inspected during the 100% inspection phase without finding i consecutive units free of defects. This value is designated S and tabled in MIL-STD-1235C by AQL indices and f values (Table V-B page 104-5). These tables are available in 1235C and should be consulted for use with section 2.4.5. Verification sampling as a check on ineffective screening is an additional feature. An operation schematic for CSP-T is shown in Figure 2.9, based on MIL-STD-1235C (Figure 5-A, page 104-3).

2.2.5 CSP-F

References to work on finite length production runs in section 2.1 include papers by Brugger (1972b) and Blackwell (1977). CSP-F is an application of these principles, a single-level continuous sampling plan similar to CSP-1, but applied to a specified number of units N to be produced in the production period considered. It is incorporated in MIL-STD-1235C and permits smaller clearance intervals i for given AOQLs and f values. MIL-STD-1235C gives tables of i and f for a wide range of N values for AQL indices and designated AOQLs (Table III-A pages 102-4 to 102-17). It uses the table of S of CSP-1 for application of the stopping or warning rule on 100% inspection. These tables should be consulted for use with section 2.4.6. It also includes verification sampling as a check on ineffective screening. The operation schematic for CSP-F is the same as CSP-1 in Figure 2.4.

2.2.6 CSP-V

In situations where there is no advantage to reducing the sampling frequency upon demonstration of good product quality (as is done on multilevel plans), reduced inspection can be achieved by using a smaller clearance interval. This is the main feature of CSP-V, also incorporated in MIL-STD-1235C. As such it is a single-level continuous sampling procedure with parameters i, f, and x, the latter being the reduced clearing interval invoked upon finding a defective unit *after* qualifying for reduced inspection. Qualification is achieved by finding i consecutive sample units free of defects on initial sampling following a clearing interval of i consecutive units on 100% inspection. Tables of f, i, and x for AQL indices and designated AOQLs are given in MIL-STD-1235C (Table VI-A, page 105-4). In this table x is always *one-third* of i and could be dropped as a distinct parameter (simply taken as one-third i). However, it is entirely feasible to use other values (and relationships) for x in devising other CSP-V sampling plans (beyond those given in MIL-STD-1235C). An associated table of S values for application of the stopping or warning

19

Figure 2.8
Operation Schematic for CSP-M

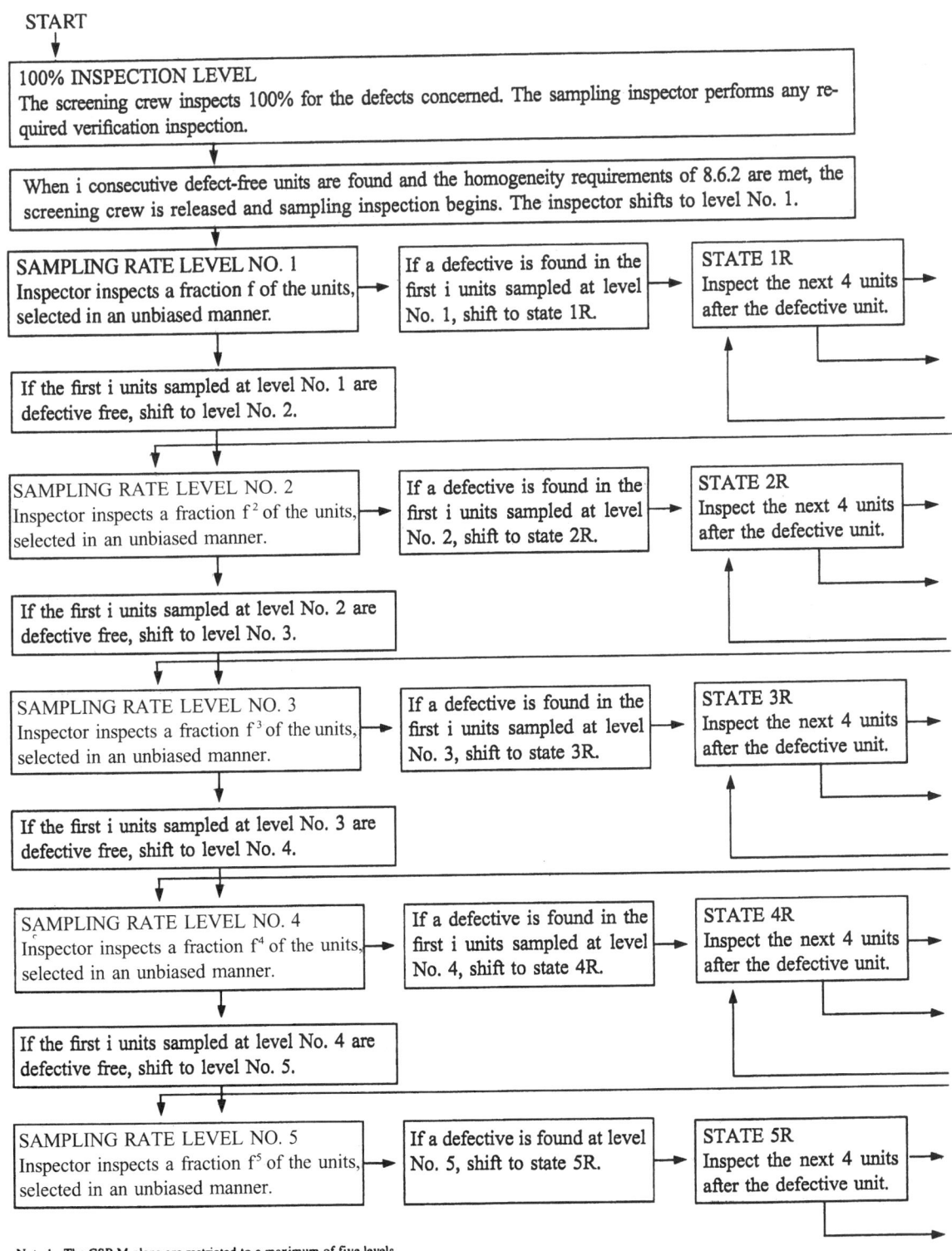

Note 1. The CSP-M plans are restricted to a maximum of five levels.
Note 2. Continue sampling at the maximum level permitted by plan used as long as it qualifies or requalifies for the next higher number level.
Note 3. Whenever the requirements of 8.6.2 are not being met, the inspector will go to 100% inspection.

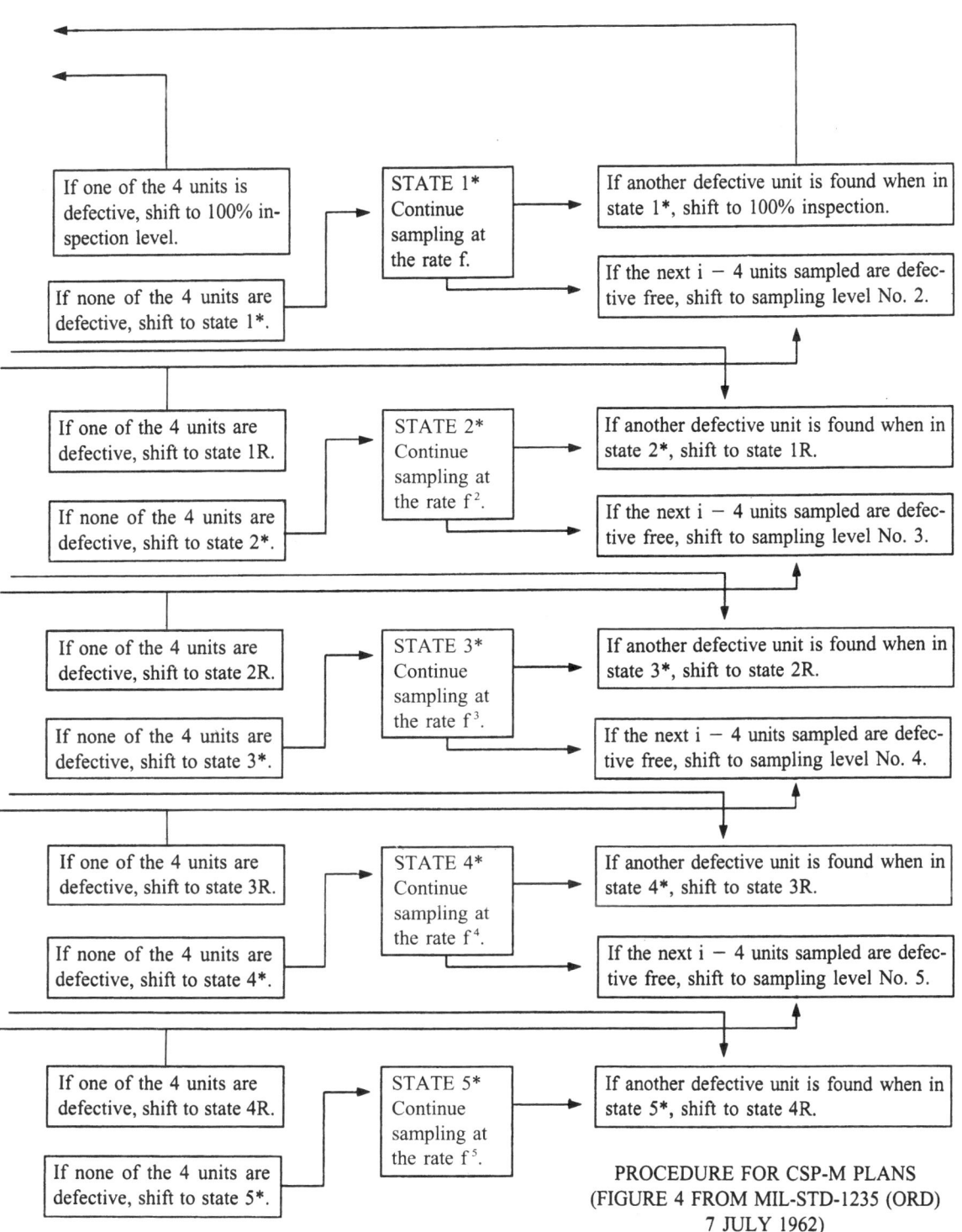

PROCEDURE FOR CSP-M PLANS
(FIGURE 4 FROM MIL-STD-1235 (ORD)
7 JULY 1962)

Figure 2.9
Operation Schematic for CSP-T

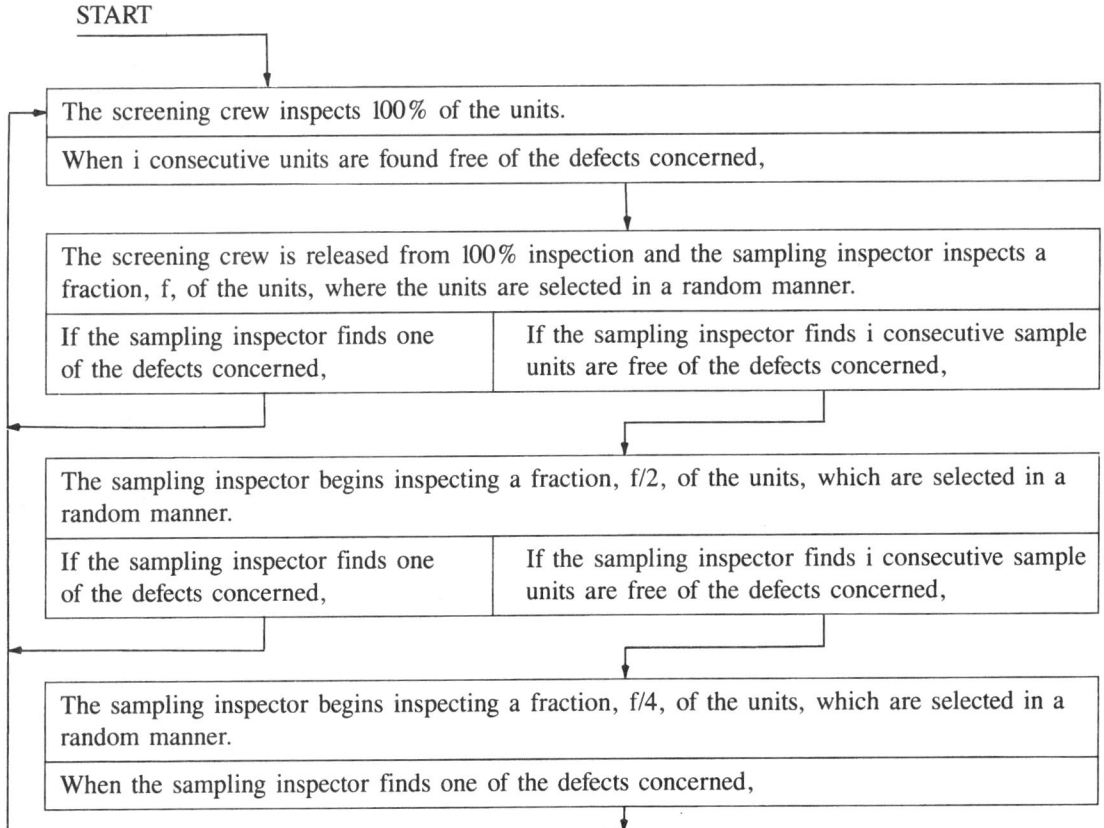

rule on 100% inspection is also given (Table VI-B, page 105-5). These tables should be consulted for use with section 2.4.7. Verification sampling as a check on ineffective screening is included in the CSP-V procedure of MIL-STD-1235C. An operation schematic for CSP-V is shown in Figure 2.10 from MIL-STD-1235C (Figure 6-A, page 105-3).

2.2.7 CSP-R

Kelley and Abraham (1969) describe a continuous sampling procedure designed to embody the normal tightened-reduced concept of MIL-STD-105E (attribute sampling). Expected inclusion of this procedure in MIL-STD-1235A did not materialize but CSP-T and CSP-V incorporate some of its features. CSP-R is also discussed in the review by Banzhaf and Brugger (1970). Tables for selecting parameters for this procedure are not available. A nomograph for CSP-R is given by Abraham (1971), but lacks correctness with respect to the parameters i and i^*.

2.2.8 CSP-C

A new development in continuous sampling plans is presented by Kandasamy and Govindaraju (1991). They refer to the plan as a "general continuous sampling plan (GCS Plan)," but this term is seen to be too broad for the procedure proposed. For simplicity, and in keeping with the earlier tradition of naming continuous sampling plans with CSP followed by a hyphenated symbol, I refer to their scheme as CSP-C, where C reflects the gist of their development in applying a conventional acceptance number (traditionally denoted by c) during the sampling phase of the CSP-1 procedure. CSP-1 is, in fact, a special case of this procedure when $c = 0$. One could be tempted to use the symbol CSP – (c + 1) so that with

Figure 2.10
Operation Schematic for CSP-V

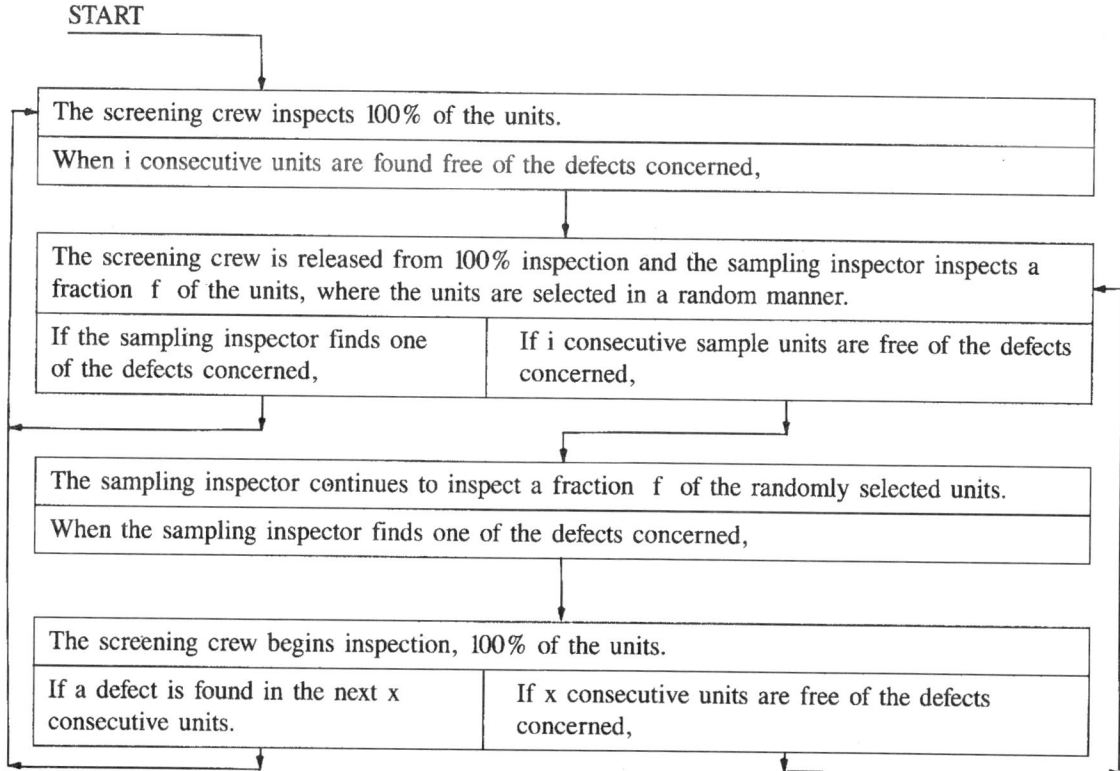

$c = 0$ this becomes CSP-1; however, it is *not* the case that for $c = 1$ or $c = 2$ we would have CSP-2 or CSP-3, respectively. Hence, as for other plans, the naming convention is symbolic and this is reflected by using capital C rather than the lowercase c, denoting the actual value of the acceptance number.

The operation schematic for CSP-C is shown in Figure 2.11. As for CSP-1 in Figure 2.4, not explicit in this schematic is the option to correct or replace defective units with good units *or* to simply remove defective units (nonreplacement option).

For the interested reader the evaluation of this procedure as carried out by Kandasamy and Govindaraju (1991) using the Markov chain development similar to Appendix A of this how-to guide is reproduced here. This should enable the development of some specific plans for applications.

As for CSP-1, the five most common measures of performance are

1. The average number of units inspected in a 100% screening sequence following the finding of a defective unit u, namely,

$$u = \frac{1 + q^i}{pq^i} \quad (2\text{-}20)$$

2. The average number of units passed under the sampling procedure before $(c + 1)$ defective units are found, v, namely,

$$v = \frac{c + 1}{fp} \quad (2\text{-}21)$$

Figure 2.11
Operation Schematic for CSP-C

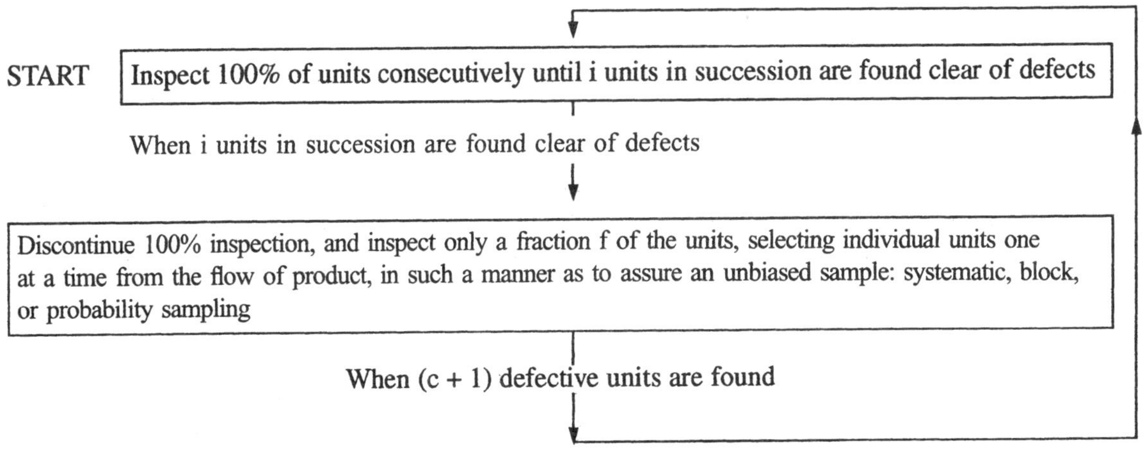

```
                    ┌─────────────────────────────────────────────────────────────────────────────────┐
                    ▼                                                                                  │
START  │ Inspect 100% of units consecutively until i units in succession are found clear of defects │  │
                                                                                                       │
       When i units in succession are found clear of defects                                           │
                              ▼                                                                        │
┌──────────────────────────────────────────────────────────────────────────────────────────────┐      │
│ Discontinue 100% inspection, and inspect only a fraction f of the units, selecting individual│      │
│ units one at a time from the flow of product, in such a manner as to assure an unbiased      │      │
│ sample: systematic, block, or probability sampling                                            │      │
└──────────────────────────────────────────────────────────────────────────────────────────────┘      │
                              ▼                                                                        │
            When (c + 1) defective units are found                                                     │
                              ▼                                                                        │
                              └────────────────────────────────────────────────────────────────────────┘
```

3. The average fraction of total produced units inspected in the long run, F, namely,

$$F = \frac{f(1 + cq^i)}{f + q^i(c + 1 - f)} \tag{2-22}$$

4. The average outgoing quality, AOQ, namely,

$$AOQ = \frac{pq^i(c + 1)(1 - f)}{f + q^i(c + 1 - f)} \tag{2-23}$$

5. The average fraction of total production accepted (passed) on a sampling basis, Pa, namely,

$$Pa = \frac{(c + 1)q^i}{f + q^i(c + 1 - f)} \tag{2-24}$$

It can be seen that for $c = 0$ these measures of performance are the same as those for CSP-1 in section 2.1.

The referenced research report by Kandasamy and Govindaraju (1991) presents a very limited table of ten CSP-C plans (one for each c value from $c = 0$ to 9). It is recommended that the authors (or others) should study and expand the application of this new continuous sampling plan procedure with tables and nomographs for practical ranges of the parameters.

2.3 NOMOGRAPHS AND THEIR USE

For the earlier CSP procedures, namely, CSP-1 and CSP-2, convenience in the selection of parameters was provided by nomographs covering a wide range of AOQL f and i values (and $k = i$ values for CSP-2). These are contained in the original papers by Dodge (1943, 1947), Dodge and Torrey (1951b), and in a later paper by Dodge (1970).

The nomograph from these papers, which combines the selection of parameters for CSP-1 and CSP-2, is shown in Figure 2.12.

A wide range of AOQLs is provided on this nomograph, ranging from 0.1% to 10%, with curves drawn for 12 specific values for each of CSP-1 and CSP-2. For any of the AOQL curves a large number of f and i combinations can be selected that will result in a plan with the given AOQL. To find intermediate or larger values of i for a given f, compute from $i_1/i_2 = AOQL_2/AOQL_1$, approximately. Further aid in the selection of the parameters f and i is provided by the nomograph in the form of p_t values.* These

---

*p_t for continuous sampling plans is defined as "the value of percent defective, in a consecutive run of $N = 1000$ product units for which the probability of acceptance, Pa, (on sampling) is 0.10 for a sample size of 100 f percent."

Figure 2.12
Nomograph for Determining Values of f and i (with k = i) for a Given AOQL for CSP-1 and CSP-2

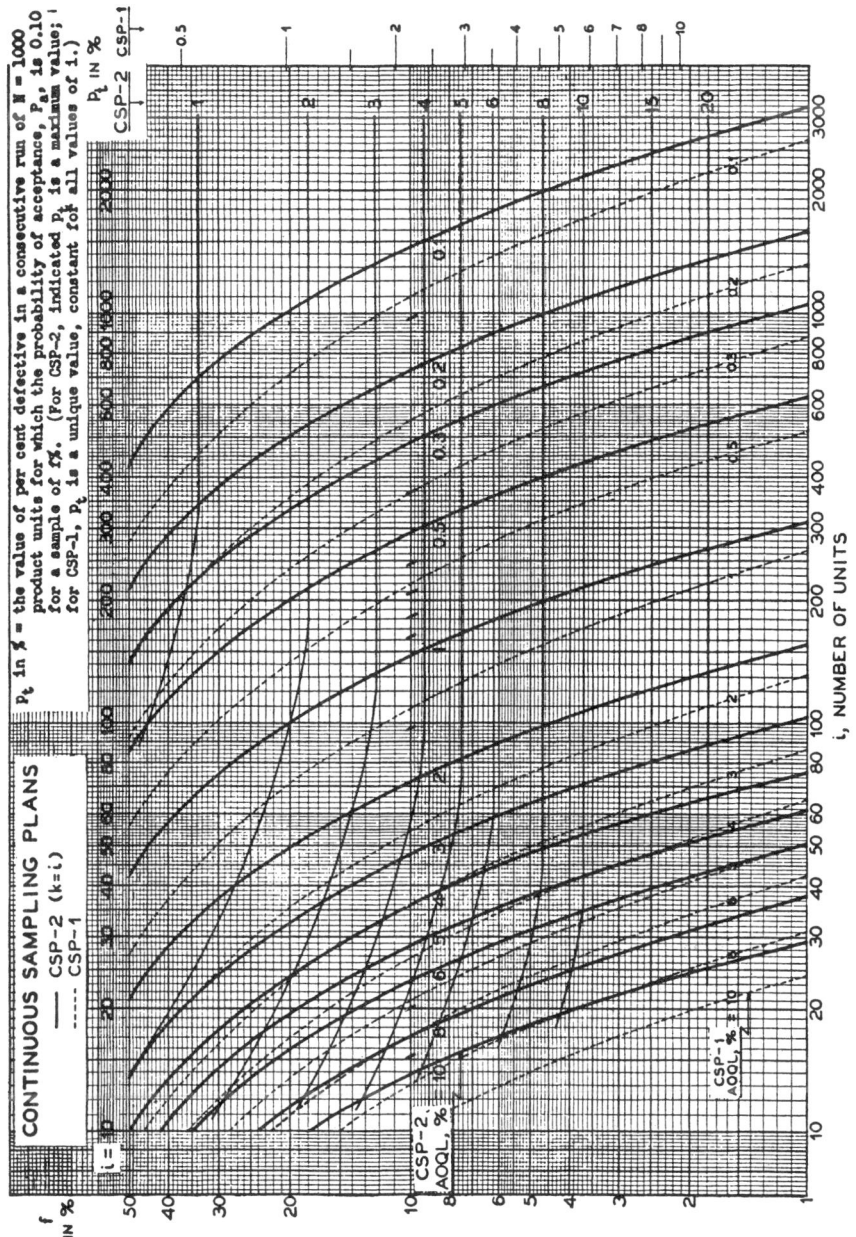

values on the right-hand scale, shown in percent defective, with corresponding straight lines for CSP-1 and curved lines for CSP-2, are associated with a level of poor quality for which the probability of acceptance *during sampling* will be small, of the magnitude of 0.10. This measure was introduced by Dodge (1943) as a control on spotty quality such as the clustering of defectives occurring during the sampling phase of continuous sampling. Hence it is not at all similar to the LQL value that might be

associated with the operating characteristic of CSP as measured by the Pa, average fraction of total production accepted on a sampling basis. This latter measure, Pa of CSP, is influenced by or dependent on i as well as f. The p_t values are influenced by or functions of f and not i. For example, acceptance during sampling occurs in the following two ways for CSP-1.
1. the unit is sampled (with probability f) and is found on inspection to be good (with probability $q = 1 - p$).
2. the unit is not sampled (with probability $1 - f$) and hence not inspected.

Therefore, the probability of acceptance during sampling for CSP-1 is

$$Pa \text{ (per unit)} = fq + (1 - f) = 1 - f(1 - q) = 1 - fp \quad (2\text{-}25)$$

Hence, Pa (over 1000 units) $= (1 - fp)^{1000}$, and p_t is that value of p (in fraction defective) for which Pa (over 1000 units) $- 0.10$, i.e.

$$(1 - fp_t)^{1000} = 0.10 \quad (2\text{-}26)$$

$$\text{Hence, } p_t = (1 - (0.10)^{0.001}) / f = (2.2999 \times 10^{-3}) / f \quad (2\text{-}27)$$

Practical use of the p_t scale indicates the necessity to use larger values of f if it is desired to provide for plans with smaller values of p_t. This is similar to the necessity to use larger sample sizes in lot-by-lot sampling to provide greater discrimination between good and bad quality. A somewhat simpler nomograph to use specifically for CSP-1 is given in Figure 2.13, which is also a companion to Figure 2.14 explained later.

To enable the selection of f and i values for CSP-1, which prescribe a desired value of LQL (in addition to AOQL) with Pa = 0.10, i.e. the *overall* probability of acceptance for CSP-1 equal to 0.10, an additional nomograph for CSP-1 is provided in Figure 2.14. Lines associated with LQL (in percent) from 0.5% to 25% are drawn on this nomograph. For more extensive use of this nomograph in the preparation of some useful tables of CSP-1 procedures with LQL values ranging from 0.50% to 32% (under a modified Fibonnacci series), see Stephens (1981). See also the discussion in section 2.4.1 on the availability of tables of CSP-1 indexed by AQL and AOQL (with known LQL) and by LQL and AOQL (with known AQL) as presented by Govindaraju (1989). Some examples from the companion nomographs of Figures 2.13 and 2.14 are as follows (with approximate values).

f(%)	i	(From 2.13) AOQL (%)	(From 2.14) LQL (%)
5	130	1.2	4.0
10	460	0.25	1.0
10	100	1.1	4.5
10	55	2.0	8.0
20	260	0.28	1.5
20	46	1.5	8.0
25	58	1.0	6.0
50	145	0.2	2.0
50	48	0.58	6.0

The nomograph of Figure 2.12 can also be used as an approximation for determining values of f and i (with k = i) for CSP-3. Dodge and Torrey (1951b) indicate that, "k = i values are equal to those for CSP-2 when AOQL is less than 2% and are less than those for CSP-2 by no more than 2 units when AOQL = 10%. Thus the effect of using Figure 2.12 for CSP-3 for the larger values of AOQL is to give actual AOQL values slightly smaller than the charted values."

Figure 2.13
CSP-1 Nomograph with AOQL Selection and p_t

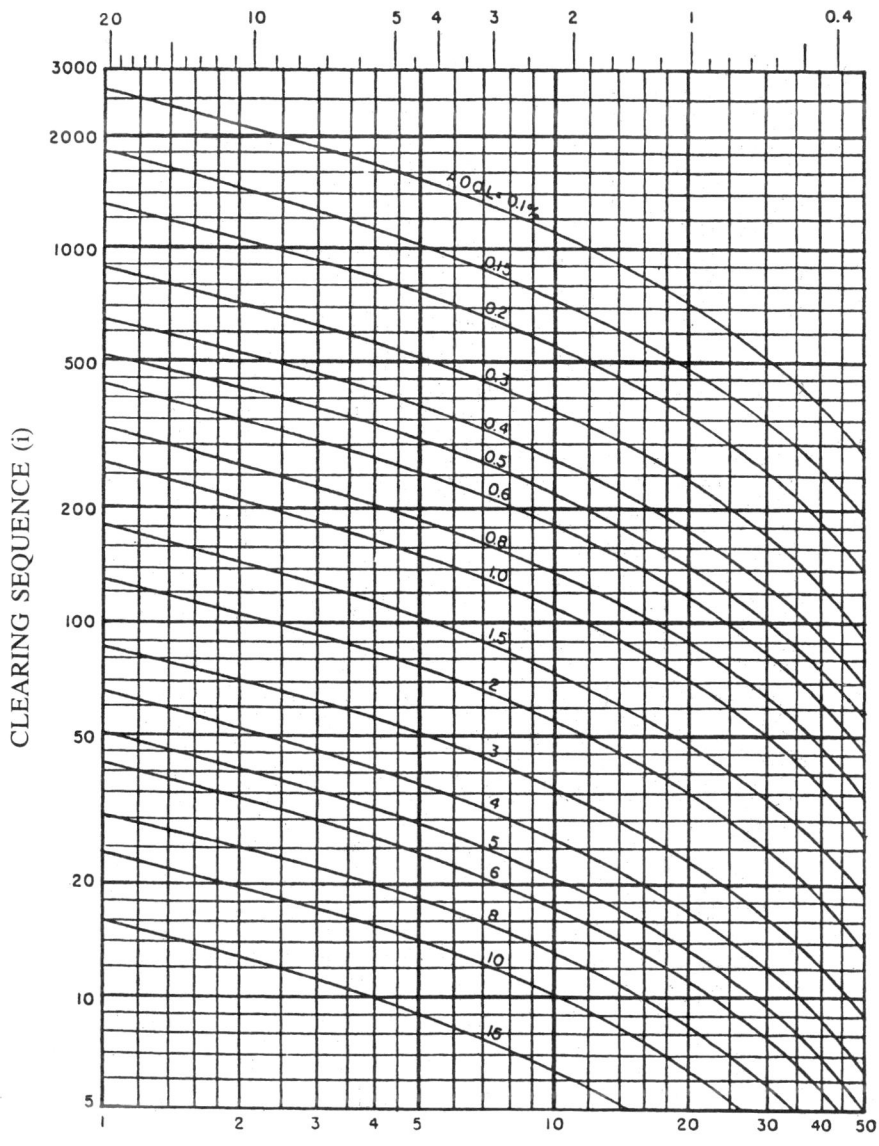

*For a run of 1000 units at a consumer's Risk of 10%.

Several examples of selection of CSP parameters from Figure 2.12 are as follows:
1. CSP-1 with AOQL = 2.0%
   a. f = 10%, i ≈ 54, with p_t ≈ 2.2%
   b. i = 100, f ≈ 2.5% (1 in 40), p_t ≈ 8.5%
   c. p_t = 3.0%, f ≈ 7.2%, i ≈ 65;
2. CSP-2 with AOQL = 4.0%
   a. f = 20%, i = k ≈ 24, p_t ≈ 3.0%

Figure 2.14
CSP-1 Nomograph with LQL Selections (Having 10% Overall Pa)

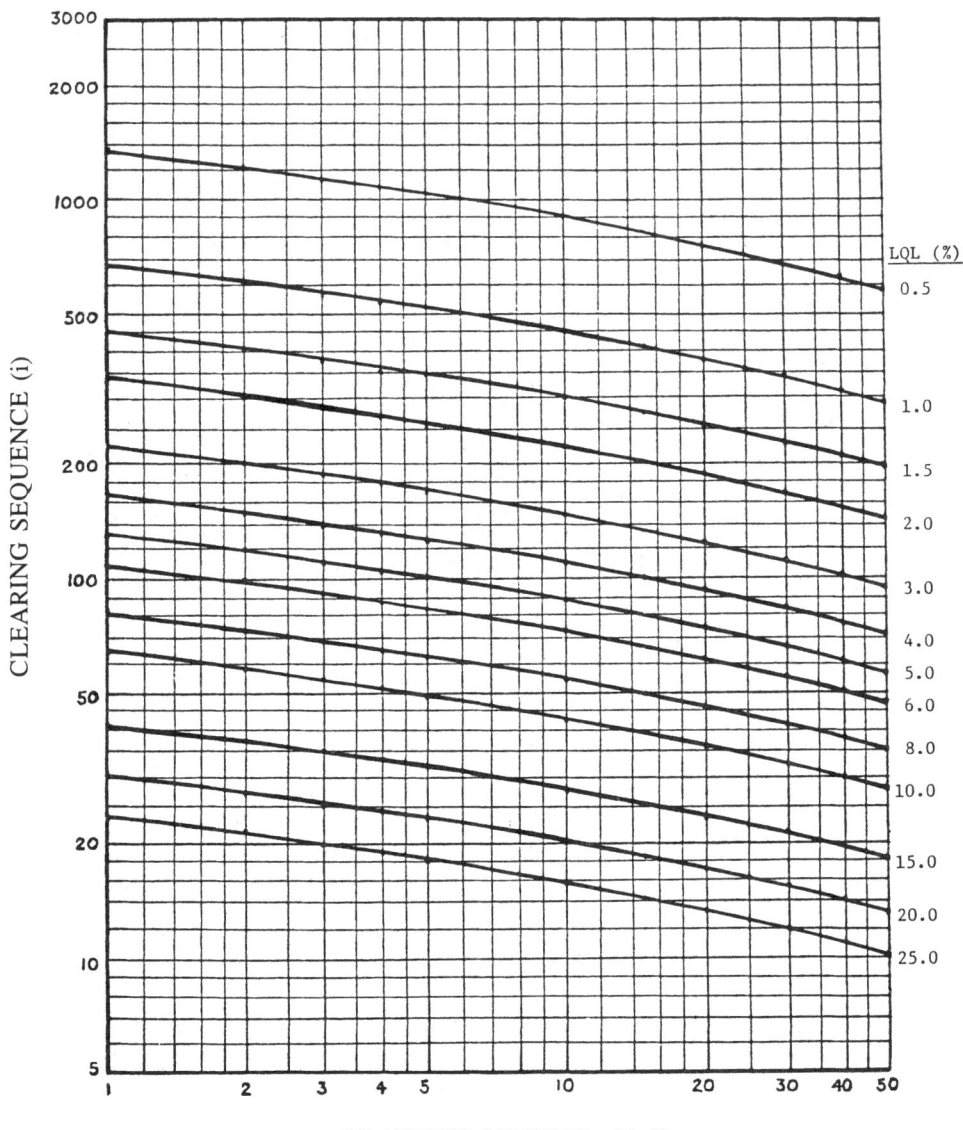

  b. $i = k = 15$, $f \approx 36\%$, $p_t \approx 2.5\%$
  c. $p_t = 5.0\%$, $f \approx 8.5\%$, $i = k \approx 38$.

Those of 2a, 2b, and 2c can be used under CSP-3 with an AOQL slightly less than 4.0%. The f value for some of the examples may be impractical to use; rounded values to whole percents or percents reflecting rational numbers (such as 14.29% representing 1 in 7) are preferred for ease of application.

For application to skip-lot sampling (see Stephens (1995)), Dodge (1955) provided a further nomograph for selecting parameters of CSP-1 plans with smaller i values. This is given in Figure 2.15 and is readily applied to continuous sampling plan situations allowing larger AOQLs.

Figure 2.15
Nomograph for Determining Values of f and i for a Given AOQL for CSP-1, Small Values of i

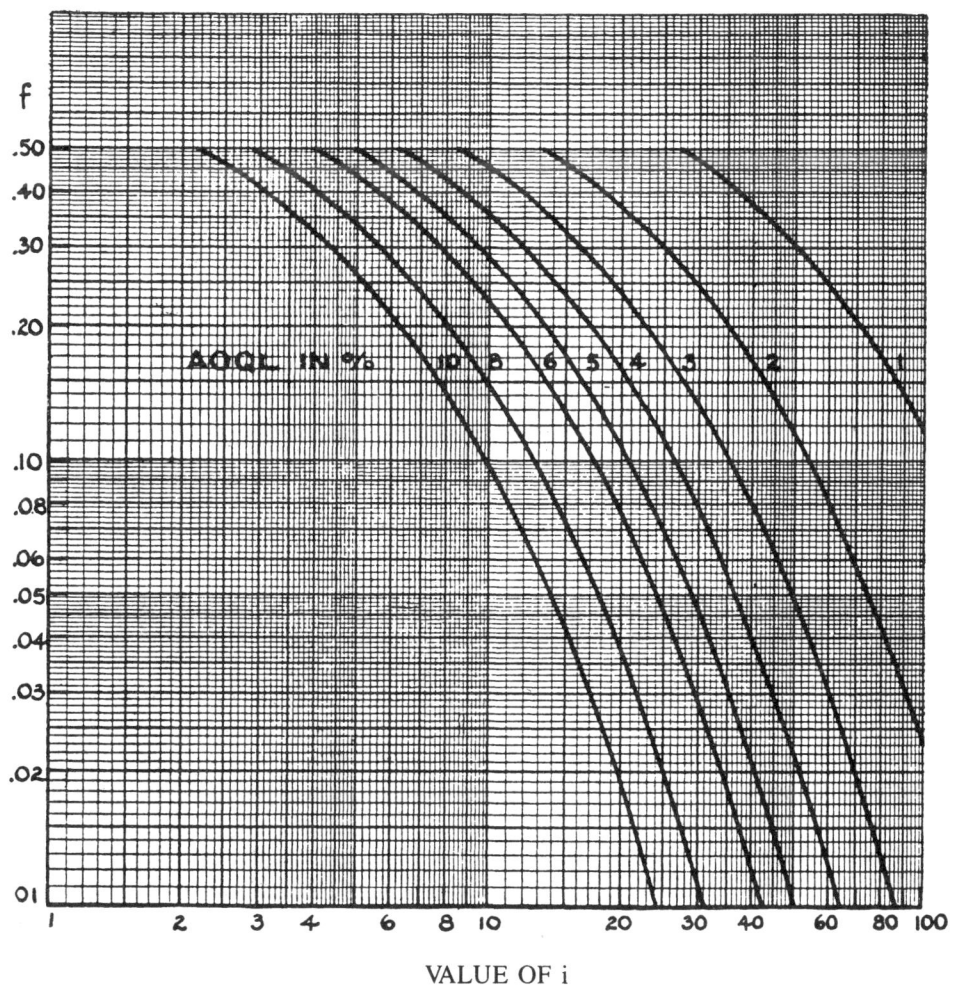

White (1961) developed an alternative nomograph for determining the AOQL, f and i parameters of CSP-1 for a fixed set of f values allowing scale selection of AOQL and i. His nomograph is given in Figure 2.16 and is useful for choosing specific f values.

Abraham (1971) presents three nomographs for continuous sampling plans, CSP-1, CSP-2, and CSP-R under a nonreplacement of defective units policy. He employs the plotting technique of White. While the nonreplacement assumption is relatively insignificant for CSP-1 (see section 2.1), Abraham's nomograph for CSP-1 extends the AOQL scale (of White) from .10% to .01% and is given in Figure 2.17. An additional nomograph for CSP-2 is given in Figure 2.18. For discussion of CSP-R see section 2.2.7.

As these nomographs clearly indicate, many combinations of f and i values will yield plans with the same AOQL. As indicated in section 2.1, there are evaluations of CSP procedures beyond the more common measures of u, v, F, AOQ, and Pa. Criteria and techniques to select the "best" of f and i values for a given AOQL have been proposed by Murphy (1958, 1959a), Resnikoff (1960), Guthrie and Johns (1958), Hillier (1964a), et al. In a subsequent paper, Murphy (1959b) developed a graphical method of determining a "best" CSP-1 plan. The technique is as follows:

1. Compute $B = -\log[(1 - A)/(1 - P)]$, where A = desired AOQL; and P = producer's nominal quality level (PNQL), a process average chosen by the producer, less than the AOQL, at which the

Figure 2.16
Alternative Nomograph for Determining Values of AOQL and i for Given f for CSP-1

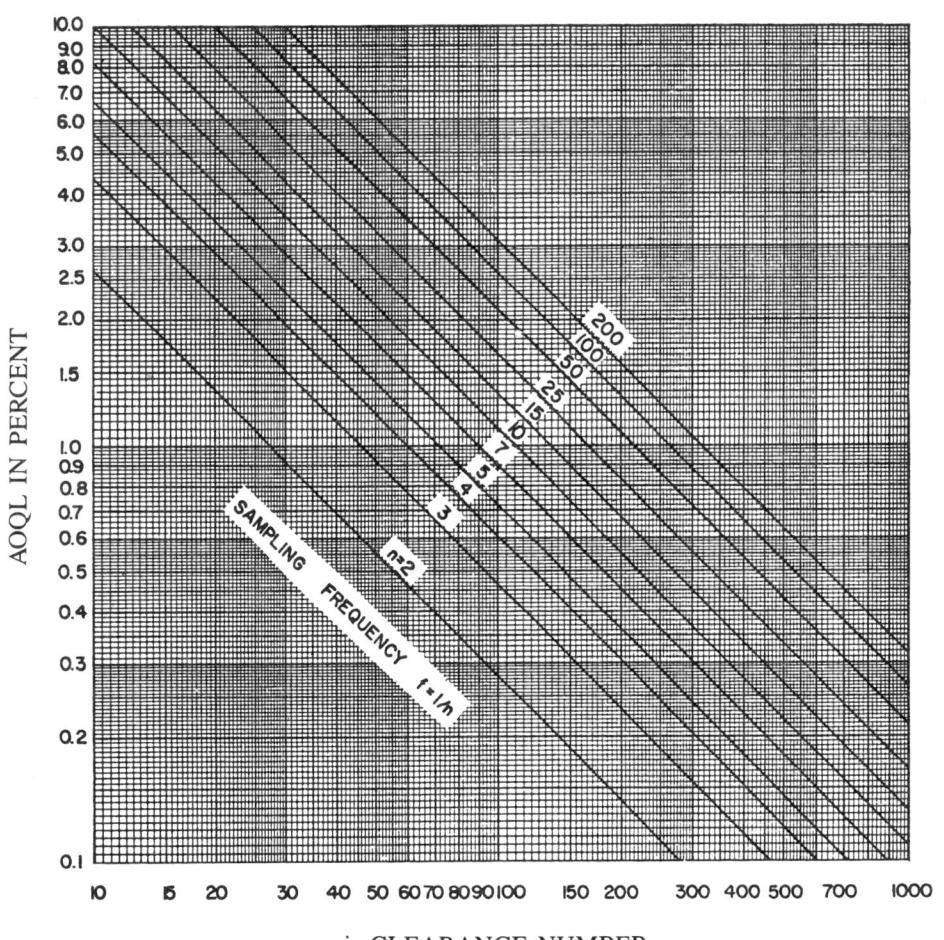

i=CLEARANCE NUMBER

producer is satisfied to operate and at which the producer desires the total fraction of units inspected (in both sampling and screening) to be a desired value F. (A, P, and F are expressed as fractions.)

2. Compute $H = [(1 - F)(1 - P)B]/(AF)$
3. Enter the abscissa of Figure 2.19 (Murphy (1959b) with H, and read the ordinate C.
4. Compute i (parameter of CSP-1) = C/B.
5. Enter Figure 2.12, 2.13, 2.15, or 2.16 with the computed i and desired AOQL A and read the corresponding value of f. (Figure 2.16 may require interpolation for f, while Figures 2.12, 2.13, and 2.15 may require interpolation to enter with A.)

For example, suppose it is desired to have F = 0.10 for an AOQL of A = 0.02, at a process average P = 0.01 (half the AOQL). Then

1. $B = -\log(.98/.99) = -\log(.989899) = 0.0044091$
2. $H = [(.90)(.99)(0.0044091)]/[(.02)(.10)] = 1.9643$
3. $C \approx 0.34$ (from Figure 2.19)
4. $i = 0.34/0.0044091 = 77.11$ or 77
5. From Figure 2.12, f = 5%

Figure 2.17
Nomograph for Determining AOQL and i of CSP-1 for a Given f, Assuming Nonreplacement

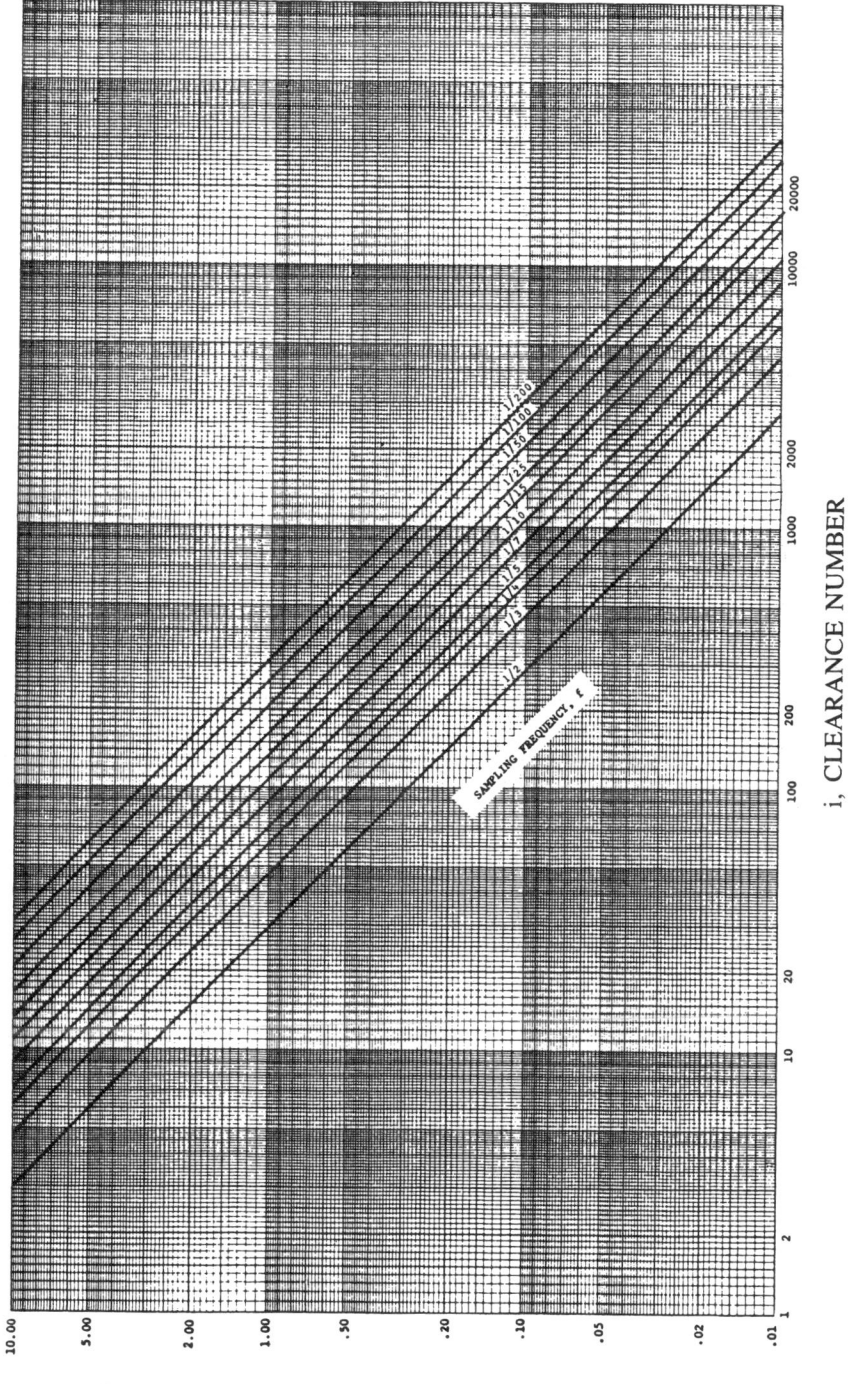

Figure 2.18
Nomograph for Determining AOQL and i (k = i of CSP-2 for a Given f Assuming Nonreplacement

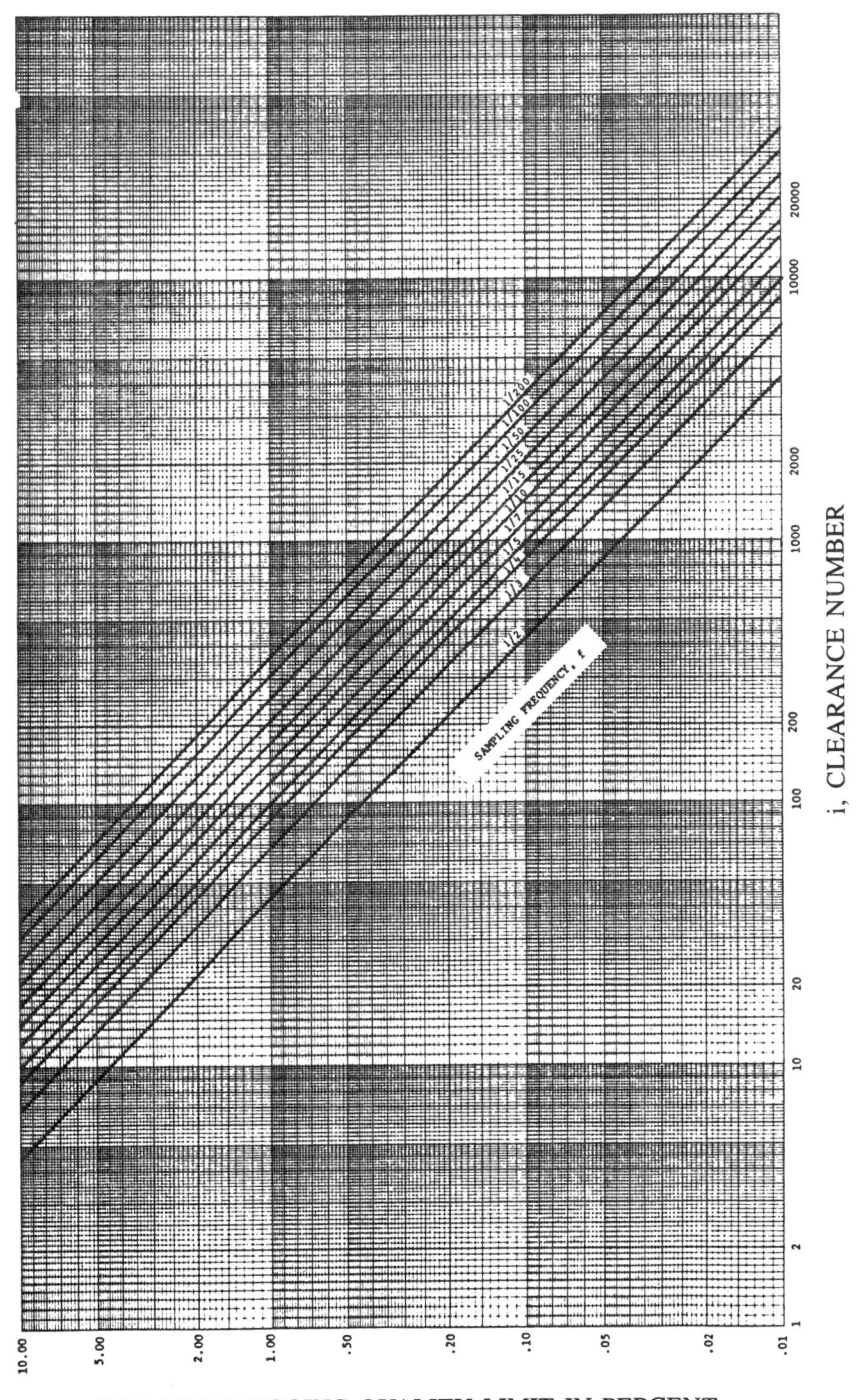

**Figure 2.19
Murphy's Chart for C, Given H**

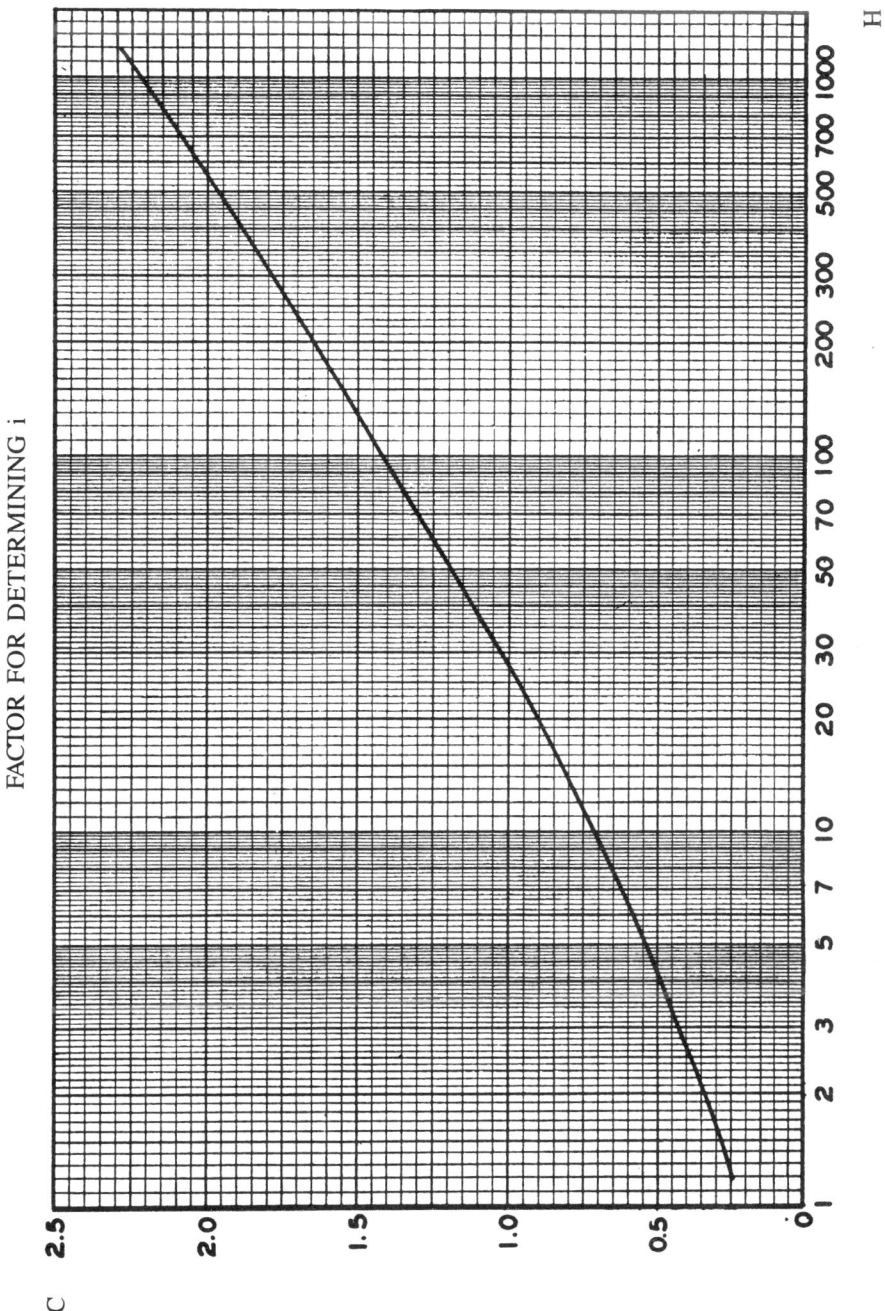

2.4 SAMPLING TABLES AND THEIR USE

Reference has been made in sections 2.1 and 2.2 to numerous tables of continuous sampling plans, including chronological development of these tables. In addition to CSP-1 and CSP-2, plans discussed in section 2.2, and especially CSP-A, CSP-M, CSP-T, CSP-F, and CSP-V, are from available tables. CSP-3 is treated as a refinement of CSP-2 and not tabled explicitly. Principal tables are MIL-STD-1234 (ORD), MIL-STD-1235A, MIL-STD-1235B, and now MIL-STD-1235C (1988), and should be procured for use with this booklet. Tables for CSP-A and CSP-M, from MIL-STD-1235 (as referenced in the discussion), are given in Appendix C and D since this standard has been revised and the earlier editions may be difficult to obtain.

2.4.1 CSP-1

While included in most tables (except those dealing exclusively with multilevel plans) the CSP-1 tables of MIL-STD-1235C are used here for selection of the parameters for these plans. Table II-A (page 101–4) gives values of i for a series of f values associated with sample frequency code letters and a series of AQL indices (serving to identify the plans only) with associated AOQLs. The sample frequency code letters are roughly associated with the number of units in the production interval as shown in Table I (page 8) of the referenced standard.

The tables also incorporate a stopping rule for the 100% screening phase, giving a table of values of S, a critical value on the number of units inspected during 100% screening without finding i consecutive conforming units, which when equaled or exceeded, signals need for corrective action on the process. The table of S values indexed by AQL and f (and sample frequency code letters) is Table II-B (page 101-5) of MIL-STD-1235C.

The operation schematic for CSP-1 is shown in Figure 2.4 of this booklet. In addition, the tables emphasize (1) homogeneity of production as a requirement for initiating sampling, (2) continued sampling by a checking inspector during periods of 100% screening as a check on ineffective screening and with the requirement that the screening crew shall start a new count of consecutive defect-free units if a defect is found by the checking inspector, and (3) use of the stopping rule on long periods of screening, with the requirement, from the tables, that "if, during a period of 100% inspection, a defect is found before finding i consecutive conforming units and after the number of units screened is equal to or greater than the appropriate value of S in Table II-B the supplier shall notify the consumer of this occurrence, and corrective action shall be taken to improve the production process. The consumer may, at its option, suspend acceptance immediately or at any time thereafter during the period of 100% inspection until the supplier corrects the cause(s) of the high rate of defectives. After effective corrective action has been taken, 100% inspection shall be reinitiated."

OC curves for the CSP-1 plans are given in Appendix A, MIL-STD-1235C (1988), along with AOQ and AFI curves.

Two examples of selection of CSP-1 parameters from Tables II-A, 1 (giving sample frequency code letters) and II-B are as follows:
1. For sampling frequency code letter C, f = 1/4 (Table II-A). For AQL = 0.4% (AOQL = 0.53%), i = 113 (Table II-A), S = 262 (Table II-B).
2. For sampling frequency code letter F, f = 1/10 (Table II-A). For AQL = 1.5% (AOQL = 1.9%), i = 57 (Table II-A), S = 221 (Table II-B).

As mentioned in section 2.3 on nomographs for CSP-1, some tables have been prepared by Govindaraju (1989) and indexed by AQL, AOQL and LQL, AOQL. The interested reader should consult this source for a limited number of CSP-1 plans with specified AQLs, LQLs, and AOQLs. The plans tabled are for the replacement option (replacement or correction of nonconforming units found). In this connection the paper contains an error with respect to the nonreplacement option. For this option the value of i found from the tables should be *increased* by one, not reduced by one as mentioned in the paper. See Appendix B for a more detailed discussion of this matter.

2.4.2 CSP-2

The table for selection of parameters for CSP-2 is also from MIL-STD-1235C, being Table IV-A (page 103-4) of the standard. It gives values of i (with k = i) for a series of f values associated with sample frequency code letters (of Table 1) and a series of AQL indices with associated AOQL values. Only the upper range of AQL values from 0.40% to 10.0% (AOQLs of 0.53% to 11.46%) are given since CSP-2 is not to be used for inspection of critical defects (generally associated with tighter AQL indices).

Values of S for the stopping rule on 100% screening for CSP-2 similar to those for CSP-1 are given in Table IV-B (page 103-5) of the standard.

The operation schematic of Figure 2.5 in this guide is applicable and the tables emphasize additional procedures as described under CSP-1. OC curves for the CSP-2 plans are also given in Appendix B of MIL-STD-1235C, as are AOQ and AFI curves.

Two examples of selection of CSP-2 parameters from Tables IV-A and IV-B (including general use of Table 1) are as follows:

1. For sampling frequency code letter B, f = 1/3 (Table IV-A). For AQL = 1.0% (AOQL = 1.22%), i = 55 (Table IV-A), S = 151 (Table IV-B).
2. For sampling frequency code letter E, f = 1/7 (Table IV-A). For AQL = 4.0% (AOQL = 4.94%), i = 25 (Table IV-A), S = 115 (Table IV-B).

2.4.3 CSP-A

As described in section 2.2.2, CSP-A is a procedure of the earlier MIL-STD-1235 (ORD) standard. Three tables from this standard are needed for selecting parameters for sampling plans. These are Table I giving AQL with its equivalent AOQL for CSP-A, Table II giving the sampling frequency code letters for CSP-A, and Table VII, the main table,* giving values of i and a for a series of f values associated with sample frequency code letters (Table II) and a series of AQL indices with associated AOQLs (agreeing with Table I).

The operation schematic for CSP-A is shown in Figure 2.7 of section 2.2.2. Additional procedures include

1. Verification sampling during the 100% screening phase with the requirement that sampling inspection shall not be instituted if any of the i consecutive units found defect-free by the screening crew are found defective by the verification inspector(s).
2. Provision for use of a tighter sampling plan upon evidence of poorer quality as specified in paragraph 8.5.5 of the referenced standard, namely "Upon resumption of inspection the consumer may, at his discretion, require the use of a sampling plan whose AQL (AOQL) is one step tighter (i.e., smaller) than normally specified if two consecutive full production intervals each had one or more inspection suspensions. Use of the tightened AQL (AOQL) will be continued until one full production interval has elapsed without inspection being suspended, at which time, the normally specified AQL (AOQL) will again be used. For critical defects, the tightened sampling plan will use a value of i twice that used in the normal sampling plan."
3. Provision to use a reduced sampling frequency upon evidence of good quality, stable, and homogeneous production, etc., as specified in paragraph 8.5.6 of the referenced standard, namely "A reduced sampling frequency equal to one-fifth of the normal sampling frequency may be used at the discretion of the consumer if all of the following conditions are satisfied:
    (a) The preceding ten full production intervals have been under normal inspection and without any inspection suspensions.

---

*Table II and Table VII have been condensed to one table in Appendix C. Table I is nonessential since specific AQLs and AOQLs are listed in the table given—the AOQLs being preferred.

(b) The estimated process average (number of defective units observed divided by number of units inspected) is less than AQL $- 3 \sqrt{AQL (1 - AQL) / (\text{number inspected})}$, where AQL is expressed as a fraction defective rather than as a percentage.

(c) Production is at a steady rate.

The normal sampling frequency shall be resumed if any one or combination of the following conditions occurs:

(a) Inspection is suspended.

(b) The estimated process average rises above the AQL.

(c) The consumer deems that the normal inspection frequency should be reinstated.

Reduced sampling frequencies shall not be used for critical defects."

OC curves for the CSP-A plans are given in the referenced standard. Since CSP-A emphasizes the incorporation of the stopping rule, a, the OC curves are in terms of "percent of time inspection will *not* be stopped" as a function of "percent defective of submitted product."

Two examples of selection of CSP-A parameters from Tables I, II, and VII (see also Appendix C) are as follows:

1. Desired AQL = 0.5%, Use AOQL = 1.08 (Table I). Units in production interval = 1000, Use code letter I' (Table II).

$$f = 1/10, i = 25, a = 3 \text{ (Table VII)}.$$

2. Desired AQL = 2.5%, Use AOQL = 3.09 (Table I or Table VII). Units in production interval = 400, Use code letter G' (Table II).

$$f = 1/10, i = 10, a = 4 \text{ (Table VII)}.$$

2.4.4 CSP-M

From section 2.2.3, CSP-M is also a procedure of MIL-STD-1235 (ORD) as is CSP-A. Tables for selecting parameters are Tables I, VIII, IX, X, XI, XII, and XIII and include AQL indexing to AOQL, i, and K values for a range of units in the production interval for $f = 1/2$ and $f = 1/3$, L values for the stopping rule on the 100% screening phase for the plans of $f = 1/2$ and $f = 1/3$, and summaries of i values for each inspection level k for $f = 1/2$ and $f = 1/3$.

The operation schematic for CSP-M is shown in Figure 2.8 of section 2.2.3. Additional procedures include

1. Verification sampling during the 100% screening phase with the requirement that sampling inspection shall not be instituted if any of the i consecutive units found defect-free by the screening crew are found defective by the verification inspector(s).

2. Use of the stopping rule during the 100% screening phase with the requirement, from the tables, that "if during a period of 100% inspection the number of units inspected exceeds the appropriate value of L in Table IX (if f equals 1/2) and Table XII (if f equals 1/3), the consumer may, at its option, suspend acceptance until the producer corrects the cause(s) of the high rate of defectives. And after the corrective action is taken, 100% inspection may be resumed."

Provision for tightened and reduced sampling is an integral part of the multilevel sampling procedure. Only the AOQ curves for the CSP-M plans are given in the referenced standard.

Two examples of selection of CSP-M parameters from the tables are as follows:

1. Desired AQL = 1.0%. Use AOQL = 1.5% (Table I or VIII).* For $f = 1/2$, units in production interval = 500, i = 43, K = 2 (Table VIII), $f_1 = 1/2$, $f_2 = 1/4$. L = 231 (Table IX).

2. Desired AQL = 0.12%. Use AOQL = 0.35% (Table I). For $f = 1/2$, units in production interval = 20,000, i = 275, K = 4 (Table VIII), $f_1 = 1/2$, $f_2 = 1/4$, $f_3 = 1/8$, $f_4 = 1/16$. L = 2090 (Table IX).

---

*Again, Table I is nonessential, since tabled values of AOQL can be selected for applications.

Only the essential tables for CSP-M are given in Appendix D. Tables VIII and IX are for f = 1/2; Tables XI and XII are for f = 1/3.

2.4.5 CSP-T

This multilevel plan described in section 2.2.4 is from MIL-STD-1235C. The simplified table for selection of parameters is given in Table V-A (page 104-4) of the standard. It gives values of i for a series of f values (first level sampling frequency) associated with sample frequency code letters (of Table I) and a series of AQL indices with associated AOQL values. As for CSP-2, only the upper range of AQL values from 0.40% to 10.0% (AOQLs of 0.53% to 11.46%) is given, since CSP-T is not to be used for inspection of critical defects.

Values of S for the stopping rule on 100% screening for CSP-T similar to those for CSP-1 and CSP-2 are given in Table V-B (page 104-5) of the standard.

The operation schematic for CSP-T is shown in Figure 2.9 of section 2.2.4. The additional procedures described for CSP-1 in section 2.4.1 are applicable, namely, homogeneity of production, check on ineffective screening, and stopping rule using the S values of Table V-B. OC, AOQ, and AFI curves are given in Appendix C of MIL-STD-1235C.

Two examples of selection of CSP-T parameters from Tables 5-A and 5-B (including general use of Table I) are as follows:
1. For sampling frequency code letter D, f = 1/5 (Table V-A), f_1 = 1/5, f_2 = 1/10, f_3 = 1/20. For AQL = 0.65% (AOQL = 0.79%), i = 106 (Table V-A). S = 320 (Table V-B).
2. For sampling frequency code letter H, f = 1/25 (Table V-A), f_1 = 1/25, f_2 = 1/50, f_3 = 1/100. For AQL = 4.0% (AOQL = 4.94%), i = 35 (Table V-A). S = 235 (Table V-B).

2.4.6 CSP-F

As described in section 2.2.5, CSP-F is similar to CSP-1 applied to finite length N production runs. And since i varies with N, extensive tables are necessary for selection of parameters. These tables are contained in MIL-STD-1235C and are given in Tables III-A-1 through III-A-12 for a range of twelve AQL indices (from 0.010% to 1.5%) and associated AOQL values (0.018% to 1.9%). Each table gives values of i for a series of f values with associated sample frequency code letters (of Table I) and a series of N values.

Values of S for the stopping rule on 100% screening for CSP-F are the same as CSP-1 as given in Table II-B. The operation schematic for CSP-F is the same as that of CSP-1 in Figure 2.4 as are the additional procedures described for CSP-1 in section 2.4.1.

OC curves and others are not provided in the referenced table supplement "since exact methods for their determination have not been developed."

Two examples of selection of CSP-F parameters from Tables III-A-1 through III-A-12 and Table II-B (including general use of Table I) are as follows.
1. For sampling frequency code letter A, f = 1/2 (Tables III-A-1 through III-A-12). For AQL = 0.10% (AOQL = 0.134%), N = 2000, i = 173 (Table III-A-6), S = 273 (Table II-B).
2. For sampling frequency code letter G, f = 1/5 (Tables III-A-1 through III-A-12). For AQL = 0.25% (AOQL = 0.33%), N = 400, i = 146 (Table III-A-8), S = 1810 (Table II-B), hence not applicable, reflecting lack of adjustment for finite production run. Complete 100% inspection is binding in this case.

2.4.7 CSP-V

As noted in section 2.2.6, CSP-V is a single-level continuous sampling procedure incorporating reduced inspection for good quality performance by way of a reduced clearing interval rather than reduced sampling frequency(ies). The reduced clearing interval x is used upon finding a defective on sampling after qualifying for reduced inspection by finding i consecutive sample units free of defects. The table for selecting parameters i and x for a series of f values with associated sample frequency code letters (of Table I) and a series of AQL indices (from 0.40% to 10.0%) with associated AOQL values (from 0.53% to

11.46%) is given in Table VI-A (page 105-4) of the standard. As mentioned earlier, note that in this table x = one-third of i for all cases.

Values of S for the stopping rule on the 100% screening for CSP-V similar to those for CSP-1 and CSP-2 are given in Table VI-B (page 105-5) of the standard.

The operation schematic for CSP-V is shown in Figure 2.10 of section 2.2.6. The additional procedures described for CSP-1 in section 2.4.1 are applicable, namely, homogeneity of production, check on ineffective screening, and stopping rule using the S values of Table VI-B.

OC, AOQ, and AFI curves are given in Appendix D of MIL-STD-1235C.

Two examples of selection of CSP-V parameters from Tables VI-A and VI-B including general use of Table 1 are as follows.

1. For sampling frequency code letter D, f = 1/5 (Table VI-A). For AQL = 0.40% (AOQL = 0.53%), i = 144, x = 48 (Table VI-A), S = 390 (Table VI-B).
2. For sampling frequency code letter A, f = 1/2 (Table VI-A). For AQL = 6.5% (AOQL = 7.12%), i = 6, x = 2 (Table VI-A), S = 13 (Table VI-B).

2.5 SUMMARY

As presented in the booklet, various CSP procedures are available for use. The following is a summary of the plans presented.

Designation	Parameters	Abstract
CSP-1	i—Clearing interval on 100% inspection f—Sampling frequency	A basic single-level continuous sampling procedure that provides for alternating between sequences of 100% inspection and sampling inspection.
CSP-2	i—Clearing interval on 100% inspection k—Clearing interval on sampling inspection f—Sampling frequency	An extension of CSP-1 that allows for sampling to continue upon finding a defect (defective), so long as the sampling defect-spacing is not too short ($\geq k$).
CSP-3	(Same as CSP-2)	A modification of CSP-2 incorporating a mini-clearing interval of the next four consecutive units, following a defect on sampling, in order for sampling to continue.
CSP-A	i—Clearing interval on 100% inspection f—Sampling frequency a—Maximum allowable number of defective units found during sequences of 100% inspection *and* sampling during a production interval.	A modification of CSP-1 allowing for a suspension of inspection (stopping rule) whenever a total of a + 1 defective units is found during all sequences of 100% inspection and sampling since the start of a production interval. It also requires use of the 100% inspection phase (requalification for sampling) at the start-up of each production interval.

CSP-M	i—Clearing interval on 100% inspection and on successive sampling levels f—Initial sampling frequency k—Maximum number of levels of exponential reduction in the sampling frequency f (k = 1, 2, 3, 4, or 5)	A multilevel continuous sampling plan, extension of CSP-3, that allows from one to five successive exponential deductions (integer powers of f) in the sampling inspection rate, each with qualification of no defects on i consecutive sampled units, including a clearing of four consecutive units, following a defect on sampling. Successive increases in the sampling rate (reduced powers of f down to the 100% inspection phase) are required when qualification fails.
CSP-T	i—Clearing interval on 100% inspection and on successive sampling levels f—Initial sampling frequency S—Maximum allowable number of units inspected during the 100% inspection phase without qualification for sampling (stopping rule)*	A tightened multilevel continuous sampling plan, which allows two successive geometric deductions (1/2, 1/4) in the sampling inspection frequency f each with qualification of no defects on i consecutive sampled units. Immediate reversion to the 100% inspection phase is required upon finding a defect at any of the sampling levels. Suspension of inspection is also featured with a stopping rule parameter S.
CSP-F	i—Clearing interval on 100% inspection f—Sampling frequency N—Specified number of units to be produced in the period considered	A single-level continuous sampling procedure, variation of CSP-1, for application to short runs of product, thereby permitting smaller clearance numbers.
CSP-V	i—Clearing interval on 100% inspection and on sampling f—Sampling frequency x—Reduced inspection clearing interval for 100% inspection	A single-level continuous sampling procedure, as an alternative to CSP-M or T, in that provision is made for reducing the clearing interval number in good quality situations where reduction of sampling frequency has no economic merit. Qualification for reduced inspection (clearing of x consecutive units on 100% inspection upon finding a defect on sampling) takes place upon finding no defects on i consecutive sampled units.
CSP-C	i—Clearing interval on 100% inspection f—Sampling frequency c—Acceptance number on sampling inspection	A generalization of CSP-1 that allows for an acceptance number other than zero (c = 0 being the special case of CSP-1) during sampling inspection.

*A stopping rule, for suspension of inspection, with parameter S, is also used with CSP-1, 2, 3, F, and V, optionally for individually designed applications, compulsorily for applications specifying use of MIL-STD-1235C.

APPENDIX A
Determination of Operating Characteristics by Markov Chains

Roberts (1965) defines the states and presents flow diagrams of transitions from the states for Continuous Sampling Plans CSP-1, CSP-2, and CSP-3. He further solves the resulting Markov Chain of CSP-1 for the equilibrium probabilities of the states and derives AOQ from these probabilities.

In the following, the Markov chains for CSP-2 and CSP-3 are solved for the equilibrium probabilities of the states. These results, together with Roberts' results for CSP-1, are then combined to derive a set of operating characteristics for CSP-1, CSP-2, and CSP-3.

A combined flow chart diagram for CSP-1, CSP-2, and CSP-3 is given by Roberts as his Figure 9 and is reproduced here with permission as Figure A.1.

From this figure the combined transition probability matrix for the Markov chains has been constructed and is shown in Figure A.2. The states have been divided into four groups, namely, the 100% Inspection States, applicable to all three plans, (2) the CSP-1 and Initial Sampling States, applicable to all three plans, and the only sampling states for CSP-1, (3) CSP-3, Rule of Four States, applicable only to CSP-3, and (4) CSP-2 Sampling States, applicable to CSP-2 and CSP-3. Since the three transition probability matrices for CSP-1, CSP-2, and CSP-3 are combined in Figure A.2, a scheme has been devised to separate the matrices for individual application.

The matrix for CSP-1 is simply that for the first $i + 4$ states with the transitions from the state Id to the states A_0 and A_1. This is emphasized by the transitions in the box numbered 1.

The matrix for CSP-2 involves these same $i + 4$ states and the $3k$ states for the CSP-2 sampling phase. The eight states for the CSP-3 rule of four are excluded. The transitions from state Id are to states Id_1, In_1 and N_1 as shown. The transitions included in the box numbered 2 are excluded from the matrix for CSP-2.

The matrix for CSP-3 involves all $i + 12 + 3k$ states. The transitions from state Id are to states d_1 and n_1. This is emphasized by the transitions in the box numbered 2.

The transition probability matrix is a stochastic matrix (its rows sum to one). Only for state Id are there multiple sets of entries, since this state serves as the link between the different procedures. The descriptions above have indicated explicitly the transitions from this state for each procedure.

CSP-1

Equations for the equilibrium probabilities of the states for CSP-1 are obtained from the successive columns of the appropriate matrix yielding $i + 4$ equations, though not an independent set. It is relatively easy to express the equilibrium probabilities for each state in terms of that for state Id, i.e., P_{Id}. Then with the additional equation due to the sum over all states equaling one, expressions for all states are obtained. They are as follows:

$$P_{A_0} = \frac{fp(1-q^i)}{D}, \text{ where } D = f + (1-f)q^i \qquad (A-1)$$

$$P_{A_J} = \frac{fpq^j}{D}, j = 1, 2, \ldots, i \qquad (A-2)$$

$$P_{Id} = \frac{fpq^i}{D} \qquad (A-3)$$

$$P_{In} = \frac{fq^{i+1}}{D} \qquad (A-4)$$

Figure A.1
Flow Chart of States and Transitions for CSP-1, CSP-2, and CSP-3

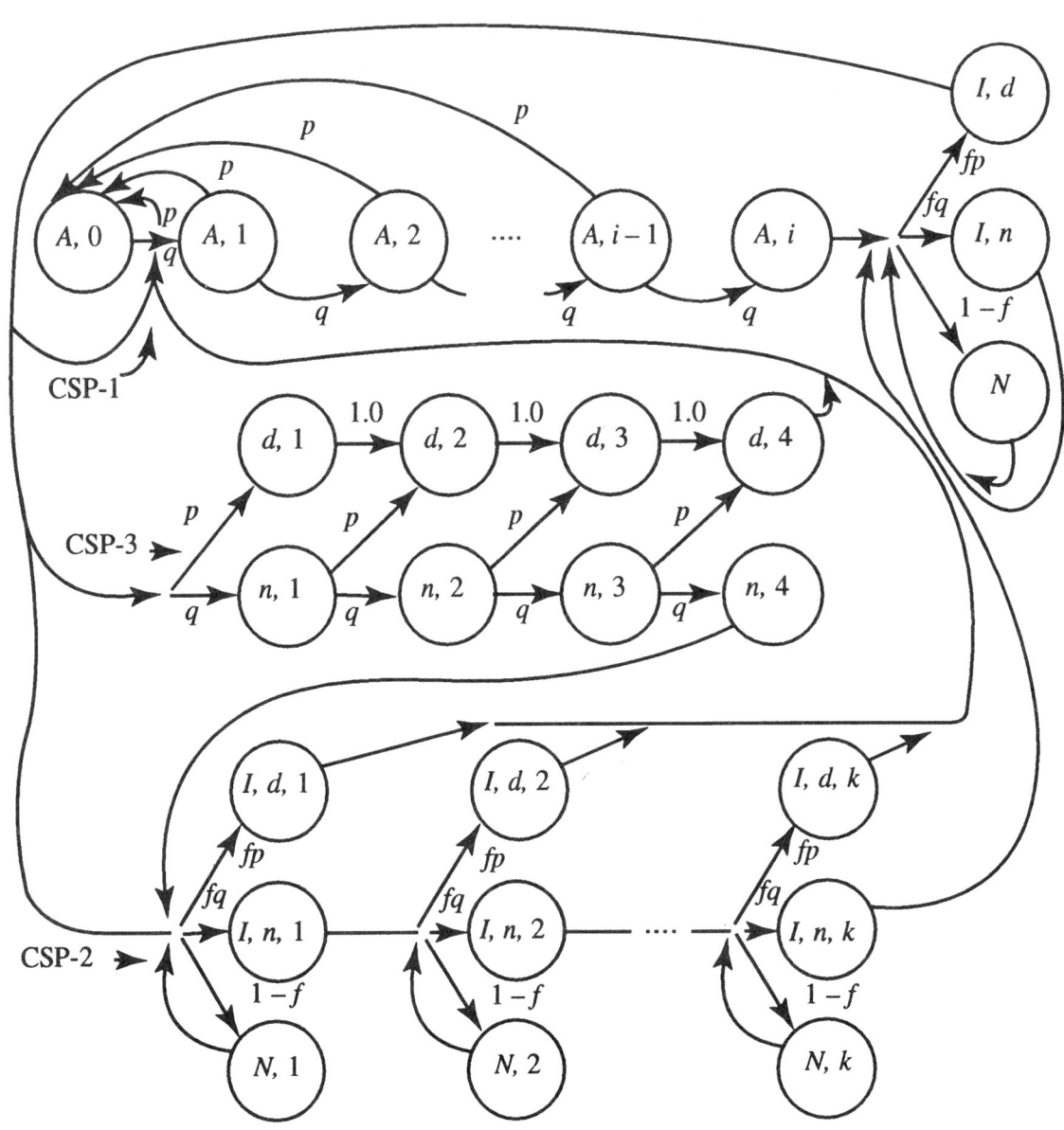

Figure A-2
Combined Transition Probability Matrices for CSP-1, CSP-2, and CSP-3

	100% Inspection States					CSP-1 (Initial Sampling States)			CSP-3, Rule of Four							CSP-2, Sampling States						
	A_0	A_1	A_2	...	A_{i-1}	A_i	Id	In	N	d_1	n_1	d_2	n_2	d_3	n_3	d_4	n_4	Id_1	In_1	N_1	Id_2	In_2	N_2	...	Id_k	In_k	N_k
A_0	p	q																									
A_1	p		q																								
A_2	p			...																							
...	p				.																						
A_{i-1}	p					q																					
A_i	p					q																					
Id							fp	fq	1−f																		
In							fp	fq	1−f																		
N							fp	fq	1−f																		
d_1										p	q																
n_1												1															
d_2												p	q														
n_2														1													
d_3														p	q												
n_3																1										
d_4																p	q										
n_4																		fp	fq	1−f						
Id_1																		fp	fq	1−f						
In_1																		fp	fq	1−f						
N_1	p					q																					
Id_2																					fp	fq	1−f			
In_2																					fp	fq	1−f			
N_2	p					q																					
...	.					.																		⋮	⋮		
Id_k																									fp	fq	1−f
In_k																									fp	fq	1−f
N_k	p					q																					

43

$$P_N = \frac{(1-f)q^i}{D} \qquad (A\text{-}5)$$

Expressions for various operating characteristics are possible by appropriate combinations of these equilibrium state probabilities.

N is the state for which the product unit is "passed over," i.e., not inspected. All states except N involve inspection of a unit, either during 100% inspection or during sampling. Hence P_N is the fraction of time (trials or units) units are not inspected. Thus if F denotes the fraction of total produced units inspected in the long run, $F = 1 - P_N$, or

$$F = 1 - \frac{(1-f)q^i}{f+(1-f)q^i} = \frac{f}{f+(1-f)q^i} \qquad (A\text{-}6)$$

The 100% inspection states are $A_0, A_1, \ldots, A_i$. Hence $\Sigma_{j=0}^{i} P_{A_j} = P_{100}$ denotes the fraction of time (trials or units) on 100% inspection. Likewise, the states Id, In, and N are the sampling states, hence $P_{Id} + P_{In} + P_N = P_s$ denotes the fraction of time (trials or units) on sampling. P_{100} and P_s are complements of unity.

$$P_s = P_{Id} + P_{In} + P_N = \frac{fpq^i}{D} + \frac{fq^{i+1}}{D} + \frac{(1-f)q^i}{D} = \frac{q^i}{f+(1-f)q^i} \qquad (A\text{-}7)$$

which measure describes the operating characteristic denoted as the probability of acceptance, P_a, for CSP-1.

A_i is the state from which sampling commences. Hence P_{A_i} is the fraction of time sampling is commenced. As denoted above, P_{100} is the fraction of time spent on 100% inspection. Hence $P_{100}/P_{A_i} = u$ denotes the average number of units inspected on 100%, namely

$$u = \frac{P_{100}}{P_{A_i}} = \frac{1-P_s}{P_{A_i}} = \frac{f(1-q^i)}{D} \frac{D}{fpq^i} = \frac{1-q^i}{pq^i} \qquad (A\text{-}8)$$

Likewise, $P_s/P_{A_i} = v$ denotes the average number of units passed under sampling, namely

$$v = \frac{P_s}{P_{A_i}} = \frac{q^i}{D} \cdot \frac{D}{fpq^i} = \frac{1}{fp}. \qquad (A\text{-}9)$$

The remaining operating characteristic of section 2.1 is AOQ, which is the fraction of units (time) not inspected, P_N, multiplied by the probability that a unit is defective, p, namely

$$\text{AOQ} = p \cdot P_N = p(1-F) = \frac{p(1-f)q^i}{f+(1-f)q^i} \qquad (A\text{-}10)$$

An interesting confirming manipulation with the state probabilities is to compute the probability that units will be inspected on sampling conditioned on being on sampling, namely

$$P\{Id \cap In | S\} = \left[\frac{fpq^i}{D} + \frac{fq^{i+1}}{D}\right] \cdot \frac{D}{q^i} = f \text{ (as expected)}$$

CSP-2

Equations for the equilibrium probabilities of the states for CSP-2 are obtained from the successive columns of the appropriate matrix as outlined above yielding $i + 4 + 3k$ equations in addition to the equation from the sum over all states equaling one. Again it is relatively easy to express the equilibrium probabilities for each state in terms of that for state Id, i.e., P_{Id}, and after obtaining P_{Id} to solve for each. They are as follows where $D = f(1-q^k)(1-q^i) + q^i(2-q^k)$.

$$P_{A_0} = \frac{fp(1-q^k)(1-q^i)}{D} \qquad (A\text{-}11)$$

$$P_{A_j} = \frac{(1 - q^k) fpq^j}{D}, \, j = 1, 2, 3, \ldots, i \tag{A-12}$$

$$P_{Id} = \frac{fpq^i}{D} \tag{A-13}$$

$$P_{In} = \frac{fq^{i+1}}{D} \tag{A-14}$$

$$P_N = \frac{(1 - f) q^i}{D} \tag{A-15}$$

$$P_{Id_n} = \frac{fp^2 q^{i+n-1}}{D}, \, n = 1, 2, 3, \ldots, k \tag{A-16}$$

$$P_{In_n} = \frac{fpq^{i+n}}{D}, \, n = 1, 2, 3, \ldots, k \tag{A-17}$$

$$P_{N_n} = \frac{(1 - f) p \, q^{i+n-1}}{D}, \, n = 1, 2, 3, \ldots, k \tag{A-18}$$

For CSP-2, the states for which the product limit is "passed over," i.e., not inspected, are N and N_n, n from 1 to k. Let $P_{FN} = P_N + \Sigma_{n=1}^k P_{N_n}$ denote the fraction of time (trials or units) units are not inspected. Hence with F as for CSP-1, $F = 1 - P_{FN}$,

$$P_{FN} = P_N + \sum_{n=1}^{k} P_{N_n} = \frac{(1 - f) q^i}{D} + \frac{(1 - f) p}{D} \sum_{n=1}^{k} q^{i+n-1} = \frac{(1 - f) q^i (2 - q^k)}{D} = 1 - F \tag{A-19}$$

and

$$F = \frac{f(1 - q^k)(1 - q^i) + fq^i (2 - p^q)}{D} \tag{A-20}$$

As for CSP-1, the 100% inspection states are $A_0, A_1, \ldots, A_i$. $P_{100} = \Sigma_{j=0}^i P_{A_j}$ denotes the fraction of time on 100% inspection, namely

$$P_{100} = \sum_{j=0}^{i} P_{A_j} = \frac{fp (1 - q^k)(1 - q^i)}{D} + \frac{(1 - q^k) fpq}{D} \sum_{j=0}^{i-1} q^j = \frac{f(1 - q^k)(1 - q^i)}{D}; \tag{A-21}$$

and the states Id, In, N and Id_n, In_n, N_n, n from 1 to k, are the sampling states so that

$$P_s = P_{Id} + P_{In} + P_N + \sum_{n=1}^{k} (P_{Id_n} + P_{In_n} + P_{N_n}) = 1 - P_{100} = \frac{q^i (2 - q^k)}{D} \tag{A-22}$$

As for CSP-1, this expresses probability of acceptance, P_a, for CSP-2. As explained for CSP-1,

$$u = \frac{P_{100}}{P_{A_i}} = \frac{f(1 - q^k)(1 - q^i)}{D} \cdot \frac{D}{(1 - q^k) fpq^i} = \frac{1 - q^i}{pq^i} \tag{A-23}$$

$$v = \frac{P_S}{P_{A_i}} = \frac{q^i (2 - q^k)}{D} \cdot \frac{D}{(1 - q^k) fpq^i} = \frac{2 - q^k}{(1 - q^k) fp} \tag{A-24}$$

and

$$AOQ = p \cdot P_{FN} = \frac{p (1 - f) q^i (2 - q^k)}{D} \tag{A-25}$$

(These results are also given in section 2.2.1.)

CSP-3

As for CSP-1 and CSP-2, equations for CSP-3 are obtained from the successive columns of the matrix of Figure A-2, yielding i + 12 + 3k equations in addition to that from the sum of the state probabilities equaling one. The solution approach through P_{ld} is again employed to obtain the following results.

$$P_{A_0} = \frac{fp(1 - q^{k+4})(1 - q^i)}{D} \tag{A-26}$$

where $D = f(1 - q^{k+4})(1 - q^i) + q^i(1 + q^4 - q^{k+4}) + 4fpq^i$

$$P_{A_j} = \frac{(1 - q^{k+4})fpq^j}{D}, j = 1, 2, 3, \ldots, i \tag{A-27}$$

$$P_{ld} = \frac{fpq_i}{D} \tag{A-28}$$

$$P_{ln} = \frac{fq^{i+1}}{D} \tag{A-29}$$

$$P_N = \frac{(1 - f)q^i}{D} \tag{A-30}$$

$$P_{d_m} = \frac{fpq^i(1 - q^m)}{D}, m = 1, 2, 3, 4 \tag{A-31}$$

$$P_{n_m} = \frac{fpq^{i+m}}{D}, m = 1, 2, 3, 4 \tag{A-32}$$

$$P_{ld_n} = \frac{fp^2 q^{i+n+3}}{D}, n = 1, 2, 3, \ldots, k \tag{A-33}$$

$$P_{ln_n} = \frac{fpq^{i+n+4}}{D}, n = 1, 2, 3, \ldots, k \tag{A-34}$$

$$P_{N_n} = \frac{(1 - f)pq^{i+n+3}}{D}, n = 1, 2, 3, \ldots, k \tag{A-35}$$

As for CSP-2,

$$P_{FN} = P_N + \sum_{n=1}^{k} P_{N_n} = \frac{(1 - f)q^i}{D} + \frac{(1-f)pq^{i+4}}{D}\sum_{n=0}^{k-1} q^n$$

$$= \frac{(1 - f)q^i(1 + q^4 - q^{k+4})}{D} = 1 - F \tag{A-36}$$

and

$$F = \frac{f(1 - q^{k+4})(1 - q^i) + fq^i(1 + q^4 - q^{k+4}) + 4fpq^i}{D} \tag{A-37}$$

CSP-3 differs from CSP-2 by the addition of the "rule of four." While this occurs during sampling, it nonetheless represents an additional inspection of 4 units. Its probability of occurrence can be computed separately (as below), but for purposes of separating sampling from 100% inspection, the rule of four is thrown in with the latter. Let $P_{R4} = \Sigma_{m=1}^{4}(P_{d_m} + P_{n_m})$ denote the fraction of time (trials or units) on rule of four checking,

$$P_{R4} = \sum_{m=1}^{4} (P_{d_m} + P_{n_m}) = 4 P_{Id} = \frac{4fpq^i}{D} \tag{A-38}$$

(since $P_{d_m} + P_{n_m} = P_{Id}$, for all m, by observation of equations (A-28), (A-31) and (A-32))

$$P_{100} = \sum_{j=0}^{i} P_{A_j} = \frac{fp(1 - q^{k+4})(1 - q^i)}{D} + \frac{(1 - q^{k+4}) fpq}{D} \sum_{j=0}^{i-1} q^j$$
$$= \frac{f(1 - q^{k+4})(1 - q^i)}{D} \tag{A-39}$$

and

$$P_s = 1 - P_{R4} - P_{100} = \frac{q^i (1 + q^4 - q^{k+4})}{D} \tag{A-40}$$

As for CSP-1 and CSP-2, this expresses probability of acceptance, Pa, for CSP-3. With u defined as in section 2.2.1 to include the rule of four,

$$u = \frac{P_{R4} + P_{100}}{P_{A_i}} = \left[\frac{4fpq^i}{D} + \frac{f(1 - q^{k+4})(1 - q^i)}{D} \right] \cdot \frac{D}{(1 - q^{k+4}) fpq^i}$$
$$= \frac{f(1 - q^{k+4})(1 - q^i) + 4fpq^i}{(1 - q^{k+4}) fpq^i} \tag{A-41}$$

$$v = \frac{Ps}{P_{A_i}} = \frac{q^i (1 + q^4 - q^{k+4})}{D} \cdot \frac{D}{(1 - q^{k+4}) fpq^i} = \frac{1 + q^4 - q^{k+4}}{(1 - q^{k+4}) fp} \tag{A-42}$$

and

$$AOQ = p \cdot P_{FN} = \frac{p(1-f) q^i (1 + q^4 - q^{k+4})}{D} \tag{A-43}$$

(These results are also given in section 2.2.1.)

Most of these results agree with Dodge and Torrey (1951b), however, those for AOQ (and hence P_a) for CSP-3 reflect corrections. The latter result (equation (A-43) for AOQ) is used by Sheesley (1975) in a FORTRAN program for CSP-3. (Correction or replacement of defective units is assumed, as is probability sampling, throughout the derivation.)

APPENDIX B
Note on CSP-1 and SkSP-1 Equations and Equivalent AOQs

This appendix addresses the matter of

(1) The statement by Dodge (1943) in a footnote, "The solution given assumes correction or replacement of defective units. Where it is expedient to reject such units and not replace them, equations (19) to (22) inclusive, should be modified by replacing i by i − 1."

(2) The statement by Dodge (1955) in a footnote, "It can be shown that i should be increased by one in CSP-1 plans when defective units are removed but not replaced"—so that Procedure A1 (replacement) uses i versus i + 1 in Procedure A2 (nonreplacement).

Some concern has been expressed that these statements are contradictory—that they represent inconsistency with respect to (i − 1) versus (i + 1) in the application of CSP-1 under the two conditions of replacement and nonreplacement. The following arguments demonstrate that the statements are *not* contradictory and that the applications are correct and consistent. The basis for the arguments is that in the earlier paper, Dodge (1943), the emphasis is on producing or obtaining the *equation* for the nonreplacement case from the *equation* for the replacement case. For the latter paper, Dodge (1955), the emphasis is on optional procedures with *equivalent AOQs* (i.e. AOQ *values*).

For each of the cases of (r)eplacement and (n)onreplacement, respectively,

$$AOQ_r(i;p,F) = p(1 - F) \tag{B-1}$$

and,

$$AOQ_n(i;p,F) = p(1 - F) / (1 - pF), \tag{B-2}$$

the former by Dodge (1943), and the latter by Case, et al (1973).

These, of course, are not equivalent for a given set of parameters, and the equations are not the same!

In fact, since $0 \leq p \leq 1$, $0 \leq F \leq 1$,

and hence $0 \leq pF \leq 1$, and $0 \leq (1 - pF) \leq 1$,

$[p(1 - F)] / (1 - pF) \geq p(1 - F)$,

i.e. $AOQ_n(i) \geq AOQ_r(i)$

$$\tag{B-3}$$

This is certainly intuitive, since AOQ_r is diluted (reduced) by the addition of corrected (or replacement by good) units—as opposed to AOQ_n not benefitting from such a dilution.

Hence, for a given set of parameters, including *i*, the AOQ for the nonreplacement option will be larger, in general, than the AOQ for the replacement case. This will be noted later.

Equations

It is possible to *obtain the equation* for $AOQ_n(i)$ *from* the equation for $AOQ_r(i)$, and vice versa with reverse substitution, by replacing i in $AOQ_r(i)$ with (i − 1), as follows:

For the CSP-1 procedure, $F = f / (f + (1 - f)q^i)$, $\tag{B-4}$

for parameters *f, i*, and $q = (1 - p)$, as per Dodge (1943).

Hence, (B-1) becomes
$$AOQ_r(i) = (p(1-f)q^i) / (f + (1-f)q^i), \qquad (B-5)$$
and (B-2) becomes
$$AOQ_n(i) = (p(1-f)q^i) / (fq + (1-f)q^i) \qquad (B-6)$$

Now, replacing i by (i – 1) in the right-hand side of (B-5) yields (note that no equal signs are used)

$$(p(1-f)q^{i-1}) / (f + (1-f)q^{i-1}),$$
$$[(p(1-f)q^i)/q] / [(fq + (1-f)q^i)/q],$$
$$(p(1-f)q^i) / (fq + (1-f)q^i),$$

which is the nonreplacement $AOQ_n(i)$, i.e., *the nonreplacement AOQ equation* with parameter i. But note that these equations (for given i) are *not equal*—as shown earlier

$$AOQ_r(i) \quad \leq \quad AOQ_n(i)$$

Hence, this procedure of replacing i (in $AOQ_r(i)$) by (i – 1) has *produced the equation* for $AOQ_n(i)$. Note Dodge's reference to, ''. . . the *equations* . . . should be modified by replacing i by (i – 1).''

Equivalent AOQ

Now, $\quad AOQ_r(i;p,f) = (p(1-f)q^i) / (f + (1-f)q^i),$

and if we want to use the nonreplacement procedure to get *this same value*; for given i, f, and p

$$AOQ_n(j;p,f) = (p(1-f)q^j) / (fq + (1-f)q^j),$$

and multiplying by: ((1/q) / (1/q)), i.e., both numerator and denominator by (1/q),

$$AOQ_n(j;p,f) = (p(1-f)q^{j-1}) / (f + (1-f)q^{j-1}),$$

hence, for this *value* ($AOQ_n(j)$) to be *equal* to $AOQ_r(i)$, we must take $j - 1 = i$,
$$\text{or } j = i + 1$$

Example

Let $p = .02$, $f = 1/2$, and $i = 14$.

$$AOQ_r = [(.02)(.5)(.98)^{14}] / [.5 + (.5)(.98)^{14}]$$
$$= 0.0075364194 / 0.8768209$$
$$= 8.5951633 \times 10^{-3} \quad \approx \quad 0.008595$$

$$AOQ_n = [(.02)(.5)(.98)^{14}] / [(.5)(.98) + (.5)(.98)^{14}]$$
$$= 0.0075364194 / 0.8668209$$
$$= 8.6943206 \times 10^{-3} \quad \approx \quad 0.008694$$

and note that $AOQ_n > AOQ_r$.
Now, let $p = .02$, $f = 1/2$, and $i = 15$.

$$AOQ_n = [(.02)(.5)(.98)^{15}] / [(.5)(.98) + (.5)(.98)^{15}]$$
$$= 0.007385691 / 0.8592845$$
$$= 8.5951633 \times 10^{-3} \quad \approx \quad 0.008595,$$

the same value as for AOQ_r with i = 14.

Hence, to obtain a nonreplacement procedure that has equivalent AOQ to the replacement procedure (with same f) *increase* i (of the replacement procedure) by 1, i.e., use i + 1. This is what Dodge (1955) was doing in proposing Procedures A1 and A2, i.e., the emphasis is on procedures having *equivalent AOQs*.

APPENDIX C

CSP-A TABLE

Table VII
Values of i and a for CSP-A Plans

Sampling Frequency Code Letter	N*	f	.015		.10		.15		.25		.40		.65		1.0		1.5		2.5		4.0		6.5		10.0	
			i	a	i	a	i	a	i	a	i	a	i	a	i	a	i	a	i	a	i	a	i	a	i	a
A'	2-8	1/1	100%	0	↓		↓		↓		↓		100%	0	↓		↓		↓		↓		100%	0	↓	
B'	9-25	1/1	100%	0	↓		↓		↓		↓		100%	0	↓		↓		↓		↓		100%	2	100%	3
C'	26-65	1/5	100%	0	↓		↓		↓		25	0	25	0	↓		↓		↓		100%	1	3	2	3	3
D'	66-110	1/10	100%	0	↓		65	0	35	0	25	0	25	0	30	1	20	1	10	1	4	1	3	2	3	3
E'	111-180	1/10	105	0	80	0	50	0	25	0	15	0	15	0	20	0	20	1	8	1	6	2	4	2	4	3
F'	181-300	1/10	100	0	60	0	40	0	20	0	↓		30	1	14	1	10	1	8	2	6	3	4	3	4	5
G'	301-500	1/10	80	0	50	0	20	0	↓		40	1	20	1	15	2	10	2	8	3	8	4	6	5	4	7
H'	501-800	1/10	80	0	30	0	↓		40	1	20	1	20	2	15	3	15	3	10	4	8	6	6	8	4	11
I'	801-1300	1/10	80	0	↓		50	1	35	1	25	2	25	3	20	4	15	4	10	6	8	8	6	12	4	16
J'	1301-3200	1/15	100	0	135	1	50	1	40	2	40	3	30	4	25	5	15	5	10	7	8	11	6	17	4	24
K'	3201-8000	1/25	120	0	100	1	60	2	50	3	40	4	30	5	30	8	20	7	10	9	9	15	7	25	5	37
L'	8001-22000	1/50	160	0	80	1	90	2	80	3	50	5	40	7	30	10	25	11	15	17	9	24	7	34	5	53
M'	22001-110,000	1/100	200	0	110	3	100	4	80	5	65	7	50	13	35	18	20	14	16	20	12	32	9	57	9	108
N'	110,001-up	1/200	240	0	120	6	100	8	80	11	60	15	45	26	30	35	25	25	20	43	12	62	15	180	12	300
			0.12		0.27		0.36		0.59		0.83		1.08		1.35		2.20		3.09		4.96		7.24		10.70	
													AOQL in %													

*N. Number of Units of Product Produced in a Production Interval is from Table II, MIL-STD-1235.
Adapted from MIL-STD-1235.

APPENDIX D

CSP-M TABLES

Table VIII
Values of i and K for CSP-M Plans, where f = 1/2

N (No. of Units in Production Interval)		.10	.15	.25	.35	.50	.75	1.0	1.5	2.0	3.0	5.0	7.5	10.0
4-65	i											5	→	4
	K											1	→	2
66-135	i									→	11	→	→	6
	K									→	1	→	→	2
136-200	i								→	18	→	→	6	→
	K								→	1	→	→	2	→
201-300	i							→	25	→	→	11	→	→
	K							→	1	→	→	2	→	→
301-400	i						→	39	→	→	20	→	→	6
	K						→	1	→	→	2	→	→	3
401-500	i					→	54	→	43	31	→	15	9	→
	K					→	1	→	2	2	→	3	3	→
501-700	i				→	82	→	65	→	→	→	→	→	→
	K				→	1	→	2	→	→	→	→	→	→
701-1100	i			→	119	→	→	→	→	→	→	→	→	→
	K			→	1	→	→	→	→	→	→	→	→	→
1101-1500	i			167	→	132	88	→	55	40	26	18	11	8
	K			1	→	2	2	→	3	3	6	4	4	4
1501-2700	i		218	→	197	→	→	→	→	→	→	→	→	→
	K		1	→	2	→	→	→	→	→	→	→	→	→
2701-4000	i	421	→	→	→	→	→	→	→	→	31	→	→	→
	K	1	→	→	→	→	→	→	→	→	4	→	→	→

AOQL (In % Defective)

Lot Size		.015	.035	.065 .10	.15	.25	.40	.65	1.0	1.5	2.5	4.0	6.5	10.0
4001-5500	i K									47 4				
5501-8500	i K		446 2	269 2				83 3	63 4	→				
8501-10500	i K	675 2	→	→	241 3	168 3	112 3	→	→					
10501-15000	i K	→	→	337 3	→	→	→	95 4	→					
15001-21000	i K	→	564 3	→	275 4	193 4	128 4	→	→					
21001-32000	i K	847 3	→	386 4	→	→	→	→	→					
32001-50000	i K	→	636 4	→	→	→	→	→	→					
50001-80000	i K	969 4	→	→	→	→	→	→	→					
80001-150000	i K	→	→	→	→	→	→	→	→	→				
150001 and Over	i K	1059 5	706 5	422 5	302 5	210 5	140 5	104 5	69 5	51 5	34 5	20 5	13 5	9 5
		.015	.035	.065 .10	.15	.25	.40	.65	1.0	1.5	2.5	4.0	6.5	10.0

AQL in %

Use the sampling below, or at the point of, the arrow. When the value of "i" equals or exceeds N, every unit must be inspected. (In such cases, sampling in accordance with MIL-STD-105 may be preferable.)

Table IX L Values for CSP-M Plans where f = 1/2

AOQL (in percent defective)	K = 1	K = 2	K = 3	K = 4	K = 5	AQL (in percent defective)
.10	1520	3395	4252	7364	9955	.015
.15	1014	2252	2837	4840	6650	.035
.25	626	1367	1705	2945	3979	.065, .10
.35	446	1005	1222	2090	2842	.15
.50	301	688	860	1480	2024	.25
.75	216	460	579	975	1326	.40
1.0	150	345	434	727	978	.65
1.5	101	231	290	484	655	1.0
2.0	71	159	218	367	486	1.5
3.0	42	111	133	240	324	2.5
5.0	20	58	78	141	194	4.0
7.5	N.A.	33	48	84	125	6.5
10.0	N.A.	23	33	62	93	10.0

NA — No Sampling Plan Given.

Table XI
Values of i and K for CSP-M Plans, where f = 1/3

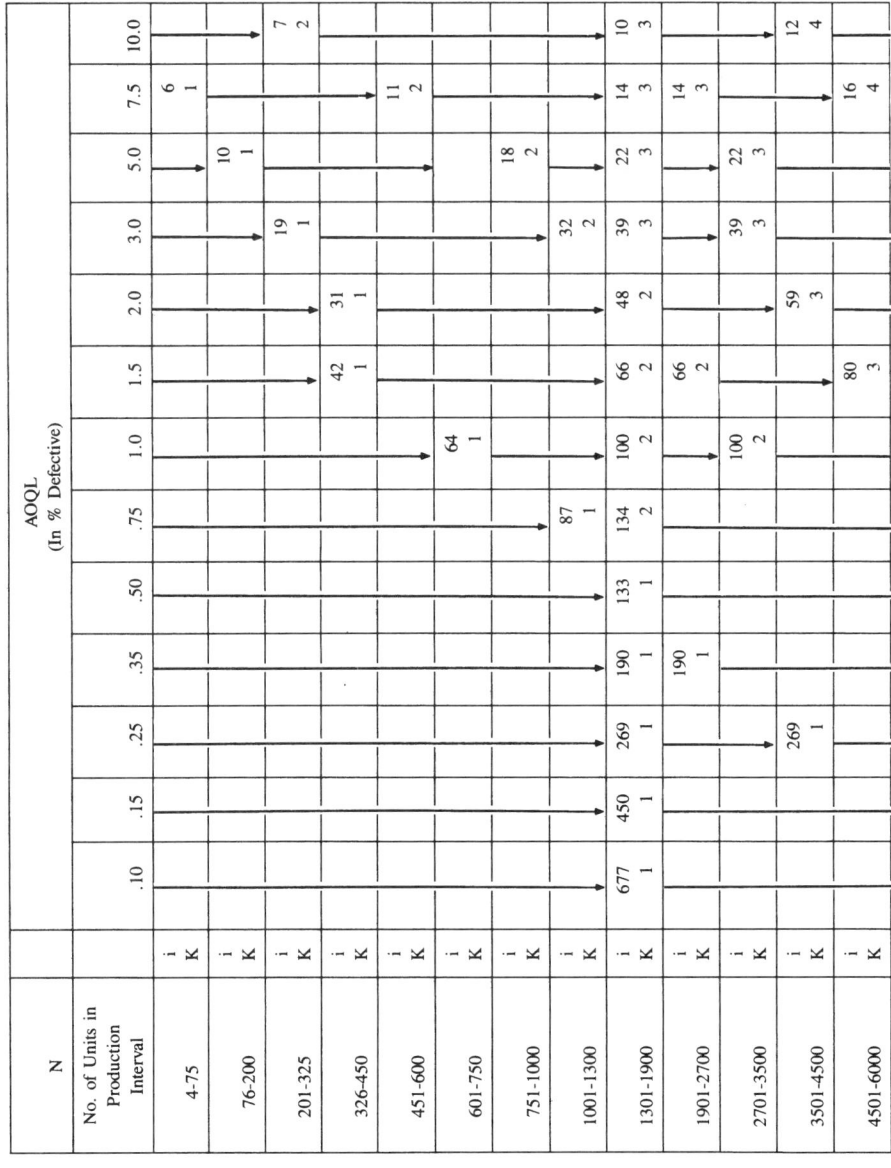

N No. of Units in Production Interval		AOQL (In % Defective)													
		.10	.15	.25	.35	.50	.75	1.0	1.5	2.0	3.0	5.0	7.5	10.0
4-75	i K												6 1	
76-200	i K											10 1		
201-325	i K										19 1		11 2	7 2
326-450	i K								42 1	31 1				
451-600	i K													
601-750	i K							64 1				18 2		
751-1000	i K						87 1				32 2			
1001-1300	i K	677 1	450 1	269 1	190 1	133 1	134 2	100 2	66 2	48 2	39 3	22 3	14 3	10 3
1301-1900	i K				190 1				66 2				14 3	
1901-2700	i K							100 2		59 3	39 3	22 3		
2701-3500	i K			269 1					80 3					
3501-4500	i K												16 4	12 4
4501-6000	i K													

Lot Size		.015	.035	.065 / .10	.15	.25	.40	.65	1.0	1.5	2.5	4.0	6.5	10.0
6001-7500	i / K	→	→	→	→	→	→	→	→	→	→	25 / 4	18 / 5	13 / 5
7501-9000	i / K	→	→	→	→	→	→	→	→	→	→	→	→	→
9001-12000	i / K	677 / 1	→	→	→	→	→	→	→	→	43 / 4	→	→	→
12001-15000	i / K	→	450 / 1	→	→	→	→	→	→	66 / 4	→	→	→	→
15001-22000	i / K	1022 / 2	680 / 2	406 / 2	290 / 2	→	161 / 3	134 / 4	89 / 4	71 / 5	46 / 5	27 / 5	18 / 5	13 / 5
22001-30000	i / K	1022 / 2	680 / 2	→	→	349 / 3	243 / 3	134 / 4	→	→	→	→	→	→
30001-40000	i / K	→	→	488 / 3	349 / 3	→	179 / 4	→	→	→	→	→	→	→
40001-55000	i / K	→	815 / 3	→	→	269 / 4	→	→	→	→	→	→	→	→
55001-75000	i / K	1224 / 3	→	540 / 4	386 / 4	→	→	→	→	→	→	→	→	→
75001-120000	i / K	→	903 / 4	→	→	→	→	→	→	→	→	→	→	→
120001-200000	i / K	1354 / 4	→	→	→	→	→	→	→	→	→	→	→	→
200001-350000	i / K	→	→	→	→	→	→	→	→	→	→	→	→	→
350001 and Over	i / K	1443 / 5	960 / 5	576 / 5	408 / 5	287 / 5	191 / 5	142 / 5	95 / 5	71 / 5	46 / 5	27 / 5	18 / 5	13 / 5
AQL in %		.015	.035	.065 / .10	.15	.25	.40	.65	1.0	1.5	2.5	4.0	6.5	10.0

Use the sampling below, or at the point of, the arrow. When the value of "i" equals or exceeds N, every unit must be inspected. (In such cases, sampling in accordance with MIL-STD-105 may be preferable.)

Table XII L Values for CSP-M Plans where f= 1/3

AOQL (in percent defective)	K = 1	K = 2	K = 3	K = 4	K = 5	AQL (in percent defective)
.10	3101	6030	10282	19294	36118	.015
.15	2047	4026	6870	12868	24038	.035
.25	1216	2412	4119	7706	14435	.065, .10
.35	855	1714	2935	5512	10237	.15
.50	609	1192	2114	3838	7221	.25
.75	411	800	1365	2560	4817	.40
1.0	296	607	1008	1916	3570	.65
1.5	210	390	681	1274	2375	1.0
2.0	150	291	505	950	1795	1.5
3.0	91	204	331	622	1167	2.5
5.0	54	109	192	363	693	4.0
7.5	29	72	119	241	466	6.5
10.0	NA	43	95	183	345	10.0

REFERENCES

Anon. 1982. Continuous Sampling Plan(s) (CSP), *Encyclopedia of Statistical Science,* Vol. 2, Kotz, S., Johnson N. L., and Read, C. B. (Editors), John Wiley and Sons, Inc., New York, pp. 174–175.

Aasheim, G. L. 1972. CSP-V: A Continuous Sampling Plan with a Provision for Reduced Clearance Number, *QEM 21-230-9,* U.S. Army Ammunition Procurement and Supply Agency, Quality Assurance Directorate, Quality Evaluation Division, Concepts Branch, Joliet, Illinois, April 1972.

Abraham, F. L. 1971. A Graphical Method of Parameter Selection for CSP-1, CSP-2, and CSP-R Under a Nonreplacement Assumption, *Journal of Quality Technology,* Vol. 3, No. 1, January 1971, pp. 2–5.

Albrecht, L., H. Gulde, A. MacLean, and P. Thompson. 1955. Continuous Sampling at Minneapolis-Honeywell, *Industrial Quality Control,* Vol. 12, No. 3, September 1955, pp. 4–9.

AMC Manual No. 74-23. 1956. Multi-Level Continuous Sampling Acceptance Plans for Attributes, Air Materiel Command, Wright-Patterson Air Force Base, Ohio, September 1956.

Anscombe, F. J. 1958. Rectifying Inspection of a Continuous Output, *Journal of the American Statistical Association,* Vol. 53, September 1958, pp. 702–719.

ANSI/ASQC Standard A2-1987. 1987. *Terms, Symbols, and Definitions for Acceptance Sampling,* American Society for Quality Control, Milwaukee, Wisc.

ANSI/ASQC Standard S1-1987. 1987. *An Attribute Skip-Lot Sampling Program,* American Society for Quality Control, Milwaukee, Wisc.

AQAP-3. 1979. List of Sampling Schemes Used in NATO Countries, NATO International Staff-Defense Support Division.

Arp, D. L. 1974. Expected Values from a Markov Chain Model of CSP-F and the Z Transform, *ARO Report 74-1, Proceedings of the Nineteenth Conference on the Design of Experiments in Army Research, Development and Testing,* October 1974, pp. 263–296.

———. 1975. Semi Markov Chains Applied to Markov Chain Models of Continuous Sampling Plans, *ARO Report 75-2, Part 2, Proceedings of the Twentieth Conference on the Design of Experiments in Army Research, Development and Testing,* October 1975, pp. 657–739.

———. 1977. Markov and Path Dependent Processes Applied to Continuous Sampling Plans in Tandem, *Proceedings of the 22nd Conference on the Design of Experiments in Army Research, Development and Testing, ARO Report 77-1,* U.S. Army Research Office, Durham, North Carolina, pp. 187–217.

Banks, J. 1989. *Principles of Quality Control,* John Wiley & Sons, Inc., New York, N.Y., pp. 441–458.

Banzhaf, R. A., and R. M. Brugger. 1970. Review of Standards and Specifications: MIL-STD-1235 (ORD), Single and Multi-Level Continuous Sampling Procedures and Tables for Inspection by Attributes, *Journal of Quality Technology,* Vol. 2, No. 1, January 1970, pp. 41–53.

Barron, C. L. 1955. An Approximation Method for Generating OC Curves for NAVORD OSTD 81 CSP Plans, *ESM 801-0-24,* Ordnance Ammunition Command, National Industrial Operations Division, Engineering Branch, Quality Assurance Section, Statistics Unit, Joliet, Illinois, June 24, 1955.

Barron, C. L., and R. Banzhaf. 1953. Technical Memo on Continuous Sampling Plans Revision, Ordnance Ammunition Center, Inspection Division, Quality Control Branch, Statistics Section, September 29, 1953.

———. 1954. Technical Memo on Continuous Sampling Plans, (Second Revision), *ESM 401-0-18,* Ordnance Ammunition Center, Inspection Division, Quality Control Branch, Joliet, Illinois, April 14, 1954.

Barron, C., D. E. Dau, and R. L. Storer. 1953. Construction of the U and V Tables in the CSP Manual, *ESM 801-0-17,* Ordnance Ammunition Center, Inspection Division, Quality Control Branch, Statistics Section, November 27, 1953.

———. 1954. Construction of OC Curves for CSP Plans, *ESM 801-0-19,* Ordnance Ammunition Center, Inspection Division, Quality Control Branch, Statistics Section, February 9, 1954.

Barron, C., W. R. Davison, and R. L. Storer. 1954. Performance of Verifying Inspection in CSP Applications, *ESM 401-0-25,* Ordnance Ammunition Center, Inspection Division, Quality Control Branch, Statistics Section, August 2, 1954.

Barron, C., W. R. Hiatt, W. R. Davison, and R. L. Storer. 1954. CSP Supplementary Standards, *ESM 401-0-22,* Ordnance Ammunition Center, Inspection Division, Quality Control Branch, Statistics Section, April 30, 1954.

Bauer, P., and P. Hackl. 1984. Optimal Continuous Sampling Procedures, contributing chapter to: *Frontiers in Statistical Quality Control,* Second Edition, Edited by H.-J. Lenz, G. B. Wetherill, and P.-Th Wilrich, Physica-Verlag, Würzburg, Germany.

Beattie, D. W. 1962. A Continuous Acceptance Sampling Procedure Based Upon a Cumulative Sum Chart for the Number of Defectives, *Journal of the Royal Statistical Society - Series C, Applied Statistics,* Vol. 11, No. 3, November 1962, pp. 137–147.

Biedenbender, R. E. 1959. Continuous Sampling Plans, *Annual Convention Transactions,* ASQC, pp. 409–420.

Blackwell, M. T. R. 1977. The Effect of Short Production Runs on CSP-1, *Technometrics,* Vol. 19, No. 3, August 1977, pp. 259–263.

Bowker, A. 1956. Continuous Sampling Plans, *Proceedings of the Third Berkeley Symposium on Mathematical Statistics and Probability,* Vol. 5, pp. 75–86.

Breeze, J. D., and J. J. Heldt. 1982. Selecting Sampling Procedures Can Be Fun, *36th Annual Technical Conference Transactions,* ASQC, Milwaukee, Wisc., pp. 709–716.

Brugger, R. M. 1967. The UAOQL of CSP-1, CSP-2, and CSP-R Under the Nonreplacement Assumption, *QEM 21-240-9,* U.S. Army Ammunition Procurement and Supply Agency, Quality Assurance Directorate, Quality Evaluation Division, Concepts Branch, Joliet, Illinois, December 1967.

———. 1972a. A Simplification of the Markov Chain Approach to Continuous Sampling Plan Formulation, *QEM 21-230-12,* U.S. Army Ammunition Procurement and Supply Agency, Quality Assurance Directorate, Quality Evaluation Division, Concepts Branch, Joliet, Illinois, March 1972.

———. 1972b. Functional Properties of CSP-1 Applied to a Finite Length Production Run, *Proceedings of the 17th Conference on the Design of Experiments in Army Research Development and Testing, Part 2, ARO-D Report 72-2,* U.S. Army Research Office, Durham, North Carolina, pp. 929–953.

———. 1974. Responsiveness Properties of Continuous Sampling Plans, *Proceedings of the 19th Conference on the Design of Experiments in Army Research Development and Testing, ARO Report 74-1,* U.S. Army Research Office, Durham, North Carolina, pp. 233–262.

———. 1976. A Simple Method for Determining the Unrestricted Average Outgoing Quality Limit (UAOQL) of a Continuous Sampling Plan, *Proceedings of the 21st Conference on the Design of Experiments in Army Research Development and Testing, ARO Report 76-2,* U.S. Army Research Office, Durham, North Carolina, pp. 409–416.

Case, K. E., E. G. Bennett, and J. W. Schmidt. 1973. The Dodge CSP-1 Continuous Sampling Plan Under Inspection Error, *AIIE Transactions,* Vol. 5, No. 3, September 1973, pp. 193–202.

Chiu, W. K., and G. B. Wetherill. 1973. The Economic Design of Continuous Inspection Procedures: A Review Paper, *International Statistical Review,* Vol. 41, No. 3, Longman Group Ltd./GB, pp. 357–373.

Connolly, C. 1991. A Sequential Approach to Continuous Sampling, *Quality Engineering,* Vol. 3, No. 4, pp. 529–535.

Curram, J. B. 1984. The Determination of Optimum Continuous Sampling Plans, contributing chapter to: *Frontiers in Statistical Quality Control,* Second Edition, Edited by H.-J. Lenz, G. B. Wetherill, and P.-Th Wilrich, Physica-Verlag, Würzburg, Germany.

Dau, D. E. and R. L. Storer. 1952. Continuous Sampling Inspection Plan, *ESM 401-10-1,* Ordnance Ammunition Center, Inspection Division, Quality Control Branch, Statistics Section, June 9, 1952.

Derman, C. S., S. Littauer, and H. Solomon. 1957. Tightened Multi-Level Continuous Sampling Plans, *The Annals of Mathematical Statistics,* Vol. 28, No. 2, June 1957, pp. 395–404. (Also published in *Technical Report No. 28,* Applied Mathematics and Statistics Laboratory, Stanford University, Stanford, California, June 20, 1956.)

Derman, C., M. V. Johns, Jr., and G. J. Lieberman. 1959. Continuous Sampling Procedures Without Control, *The Annals of Mathematical Statistics,* Vol. 30, No. 4, December 1959, pp. 1175–1191. (Also published in *Technical Report No. 39,* Applied Mathematics and Statistics Laboratory, Stanford University, Stanford, California, October 10, 1958.)

Dodge, H. F. 1943. A Sampling Inspection Plan for Continuous Production, *The Annals of Mathematical Statistics,* Vol. 14, No. 3, September 1943, pp. 264–279 and *Transactions of the ASME,* Vol. 66, No. 2, February 1944, pp. 127–133; and also reprinted in the *Journal of Quality Technology,* Vol. 9, No. 3, pp. 104–119, July 1977.

———. 1947. Sampling Plans for Continuous Production, *Industrial Quality Control,* Vol. IV, No. 3, November 1947, pp. 5–9.

———. 1948. Administration of a Sampling Plan, *Industrial Quality Control,* Vol. V, No. 3, November 1948, pp. 12–19.

———. 1955. Skip-lot Sampling Plan, *Industrial Quality Control,* Vol. 11, No. 5, February 1955, pp. 3–5.

———. 1960. Book review of: *Inspection and Quality Control Handbook* (Interim) H106, Multi-Level Continuous Sampling Procedures and Tables for Inspection by Attributes, *Technometrics,* Vol. 2, No. 4, November 1960, pp. 518–520.

———. 1970. Notes on the Evolution of Acceptance Sampling Plans, Part IV, *Journal of Quality Technology,* Vol. 2, No. 1, January 1970, pp. 1–8.

Dodge, H. F., and H. G. Romig. 1959. *Sampling Inspection Tables—Single and Double Sampling,* John Wiley & Sons, New York, 2nd edition, 1959.

Dodge, H. F., and M. N. Torrey. 1951a. Continuous Sampling Plans, *Monograph 1834,* Bell Telephone System Technical Publications.

———. 1951b. Additional Continuous Sampling Plans, *Industrial Quality Control,* Vol. 7, No. 5, March 1951, pp. 7–12.

Duncan, A. J. 1986. *Quality Control and Industrial Statistics,* 5th edition, Richard D. Irwin, Homewood, Illinois.

Elfving, G. 1962. The AOQL of Multi-Level Continuous Sampling Plans, *Z. Wahrscheinlichkeitstheorie,* 1, Springer-Verlag, Berlin, pp. 70–81. (Also published in *Technical Report No. 48,* Applied Mathematics and Statistics Laboratory, Stanford University, Stanford, California, December 5, 1960.)

Endres, A. 1967a. A Model for Obtaining the Operating Characteristics of a Skip Lot Sampling Procedure, U.S. Army Ammunition Procurement and Supply Agency, Quality Assurance Directorate, Quality Evaluation Division, Concepts Branch, Joliet, Illinois.

———. 1967b. Derivation of the Operating Characteristic Curve Formula of Project SKIP, *QEM 21-230-3,* U.S. Army Ammunition Procurement and Supply Agency, Quality Assurance Directorate, Quality Evaluation Division, Concepts Branch, Joliet, Illinois, July 1967.

———. 1967c. The Unrestricted AOQL and Its Use in Continuous Sampling Plans, *QEM 22-240-2,* U.S. Army Ammunition Procurement and Supply Agency, Quality Assurance Directorate, Quality Evaluation Division, Concepts Branch, Joliet, Illinois, July 1967.

———. 1969. The Computation of the Unrestricted AOQL When Defective Material is Removed but not Replaced, *Journal of the American Statistical Association,* Vol. 64, No. 326, June 1969, pp. 665–668. (Also published in *QEM 21-240-8,* U.S. Army Ammunition Procurement and Supply Agency, Quality Assurance Directorate, Quality Evaluation Division, Concepts Branch, Joliet, Illinois, November 1967.

———. 1979. Minimum AFI Plan Designs for Inspection Error, *33rd Annual Technical Conference Transactions,* ASQC, Milwaukee, Wisc., May 14–16, 1979, pp. 414–416.

ESM 401-0-6a. 1953. *Instructions for Inspection by Attributes on Moving Product—Continuous Sampling Plans,* Ordnance Ammunition Center, U.S. Army, May 1953.

ESM 401-0-17. 1954. *A Manual of Procedures and Tables for Sampling Inspection by Attributes on Moving Product—Continuous Sampling Plans,* Ordnance Ammunition Center, U.S. Army, Second Revision, April 1954.

Ewan, W. D., and K. W. Kemp. 1960. Sampling Inspection of Continuous Processes With No Autocorrelation Between Successive Results, *Biometrika,* Vol. 47, Nos. 3 and 4, pp. 363–380.

Fordice, J. J. 1972. A Tightened Multi-Level Continuous Sampling Plan, CSP-T, *QEM 21-230-10,* U.S. Army Ammunition Procurement and Supply Agency, Quality Assurance Directorate, Quality Evaluation Division, Concepts Branch, Joliet, Illinois, April 1972.

Gessford, J. 1955. Mathematical Treatise to Manual on Multi-Level Continuous Sampling, Department of Industrial Engineering, Stanford University, Stanford, California, June 16, 1955.

Girshick, M. A. 1954. A Sequential Inspection Plan for Quality Control, *Technical Report No. 16,* Applied Mathematics and Statistics Laboratory, Stanford University, Stanford, California, July 23, 1956.

Girshick, M. A., and H. Rubin. 1952. A Bayes Approach to a Quality Control Model, *The Annals of Mathematical Statistics,* Vol. 23, pp. 114–125.

Godfrey, A. B., and A. B. Mundel. 1984. Guide for Selection of an Acceptance Sampling Plan, Review of Standards and Specifications, *Journal of Quality Technology,* Vol. 16, No. 1, January 1984, pp. 50–55.

Govindaraju, K. 1989. Procedures and Tables for the Selection of CSP-1 Plans, *Journal of Quality Technology,* Vol. 21, No. 1, January 1989, pp. 46–50.

Grant, E. L., and R. S. Leavenworth. 1988. *Statistical Quality Control,* 6th edition, McGraw-Hill Book Co., New York.

Gregory, G. 1956. An Economic Approach to the Choice of Continuous Sampling Plans, *Technical Report No. 30,* Applied Mathematics and Statistics Laboratory, Stanford University, Stanford, California, September 20, 1956.

———. 1957. Statistical Quality Control: A Review of Continuous Sampling Plans, *Journal of the Textile Institute,* Vol. 48, pp. 467–481.

Gurfel, B. 1981. The Stochastic Optimal Model of Continuous Production Process Control, *International Journal of Production Research,* Vol. 19, No. 5, September-October 1981, pp. 505–513.

Guthrie, D., and M. V. Johns, Jr. 1958. Alternative Sequences of Sampling Rates for Tightened Multi-Level Continuous Sampling Plans, *Technical Report No. 36,* Applied Mathematics and Statistics Laboratory, Stanford University, Stanford, California, February 18, 1958.

H-106. 1958. *Multi-Level Continuous Sampling Procedures and Tables for Inspection by Attributes,* Inspection and Quality Control Handbook (Interim), Office of the Assistant Secretary of Defense (Supply and Logistics), Washington, D.C., 31 October 1958.

H-107. 1959. *Single Level Continuous Sampling Procedures and Tables for Inspection by Attributes,* Inspection and Quality Control Handbook (Interim), Office of the Assistant Secretary of Defense (Supply and Logistics), Washington, D.C., 30 April 1959.

Hansen, B. L., and P. M. Ghare. 1987. *Quality Control Application,* Prentice-Hall, Englewood Cliffs, N.J.

Hassan, M. Zia. 1969a. (MLP-RS) Multi-Level Continuous Sampling Procedure for Inspection by Attributes, Illinois Institute of Technology, Chicago, Illinois.

———. 1969b. Multi-Level Continuous Sampling Procedure for Inspection by Attributes, *AIIE Transactions,* Vol. 1, No. 3, September 1969, pp. 257–266.

Hassan, M. Zia, and A. Endres. 1978. Manufacturing Systems Containing CSP and Lot Plans, *32nd Annual Technical Conference Transactions,* ASQC, Milwaukee, Wisconsin, May 8–10, pp. 79–84.

Heldt, J. J. 1981. Continuous Sampling Plans and Skip Lot Application, *Quality,* Hitchcock Publishers, Wheaton, Illinois, June 1981, pp. 65–66.

Hillier, F. S. 1964a. New Criteria for Selecting Continuous Sampling Plans, *Technometrics,* Vol. 6, No. 2, May 1964, pp. 161–178. (Also published in *Technical Report No. 52,* Applied Mathematics and Statistics Laboratory, Stanford University, Stanford, California, May 10, 1961.)

———. 1964b. Continuous Sampling Plans Under Destructive Testing, *Journal of the American Statistical Association,* Vol. 59, No. 306, June 1964, pp. 376–401.

Ireson, W. G. 1956. Multi-Level Continuous Sampling Acceptance Plans for Attributes, Proposed *AMC Manual 74,* Department of Industrial Engineering, Stanford University, Stanford California, 15 June 1956.

Ireson, W. G., and R. E. Biedenbender. 1958. Multi-Level Continuous Sampling Procedures and Tables for Inspection by Attributes, *Industrial Quality Control,* Vol. 15, No. 4, October 1958, pp. 10–15.

Jafri, S. Q. 1988. Skip Parts Continuous Sampling Plans for Kanban (Just-In-Time) Environment, with Three Dimension Decision Criteria, *Proceedings of the 32nd EOQC Annual Conference,* Moscow, USSR, 13–17 June 1988, pp. 120–126.

Juran, J. M. 1988. (Editor-in-Chief), *Quality Control Handbook,* 4th edition, McGraw-Hill Book Company, New York.

Kandasamy, C., and K. Govindaraju. 1991. A General Continuous Sampling Plan Having Acceptance Number, *Research Report No. 33,* Bharathiar University, Department of Statistics, Coimbatore-641 046, Tamilnadu, India, September 1991.

Kao, E. P. C. 1972. Economic Screening of a Continuously Manufactured Product, *Technometrics,* Vol. 14, No. 3, August 1972, pp. 653–661.

Kelly, H. W., and F. L. Abraham. 1967. CSP-R: A Continuous Sampling Plan with Provision for Normal, Tightened, and Reduced Inspection, *QEM 21-230-5,* U.S. Army Ammunition Procurement and Supply Agency, Quality Assurance Directorate, Quality Evaluation Division, Concepts Branch, Joliet, Illinois, September 1967.

———. 1969. Theory and Assumptions Underlying the Development of CSP-R, *Proceedings of the Fourteenth Conference on the Design of Experiments in Army Research Development and Testing,* U.S. Army Research Office, Durham, North Carolina, pp. 79–102. (Also published in *QEM 21-230-6,* U.S. Army Ammunition Procurement and Supply Agency, Quality Assurance Directorate, Quality Evaluation Division, Concepts Branch, Joliet, Illinois, November 1968.

Koopman, L. 1965. Three CSP Models with Three Levels of Inspection Severity, *QEM 21-230-1,* U.S. Army Ammunition Procurement and Supply Agency, Quality Assurance Directorate, Quality Evaluation Division, Methods Branch, March 17, 1965.

Kosik, P., and K. Sarkadi. 1984. Continuous Sampling by Sequential Method, contributing chapter to: *Frontiers in Statistical Quality Control,* Second Edition, Edited by H.-J. Lenz, G. B. Wetherill, and P.-Th Wilrich, Physica-Verlag, Würzburg, Germany.

Kumar, V. S. S. 1983. Note on MIL-STD-1235 (ORD) Continuous Sampling Procedures for Markov-Dependent Processes, *Defence Science Journal,* Vol. 33, No. 4, October 1983, pp. 309–316.

———. 1984. A Tightened *m*-level Continuous Sampling Plan for Markov-Dependent Production Processes, *AIIE Transactions,* Vol. 16, No. 3, September 1984, pp. 257–261.

———. 1985. Some Aspects of Continuous Sampling Plans for Markov-Dependent Production Processes, *Mathematische Operationforschung und Statiskik, Series Statistics,* Vol. 16, No. 4, pp. 569–576.

Kumar, V. S. S., and M. B. Rajarshi. 1987. Continuous Sampling Plans for Markov-Dependent Production Processes, *Naval Research Logistics Quarterly,* Vol. 34, pp. 629–644.

Lasater, H. A. 1970. On the Robustness of a Class of Continuous Sampling Plans Under Certain Types of Process Models, unpublished Ph.D. dissertation, Rutgers University, New Brunswick, N.J., June 1970.

LeMaster, V., and R. McKeague. 1958. Stopping Rules Used in the Proposed DoD Handbook on Single-Level Continuous Sampling Plans, *ESM 401-0-29,* Ordnance Ammunition Command, National Industrial Operations Division, Engineering Branch, Inspection Engineering and Standards Section, Methods Development Unit, Joliet, Illinois, November 28, 1958.

Lieberman, G. J. 1953. A Note on Dodge's Continuous Inspection Plan, *The Annals of Mathematical Statistics,* Vol. 24, No. 3, December 1953, pp. 480–484. (Also published in *Technical Report No. 12,* Applied Mathematics and Statistics Laboratory, Stanford University, Stanford, California, 1953.)

———. 1955. Continuous Sampling Procedures, *Proceedings of the First Annual Statistical Engineering Symposium,* Army Chemical Center, Maryland, pp. 11–21.

Lieberman, G. J., and A. H. Bowker. 1958. Recent Developments in Continuous Sampling, *Bulletin de l'Institut International de Statistique,* Stockholm.

Lieberman, G. J., and H. Solomon. 1955. Multi-Level Continuous Sampling Plans, *The Annals of Mathematical Statistics,* Vol. 26, No. 4, December 1955, pp. 686–704. (Also published in *Technical Report No. 17,* Applied Mathematics and Statistics Laboratory, Stanford University, Stanford, California, September 1954.)

Magwire, C. 1956. Finite Continuous Sampling Plans Involving a Stopping Rule, *Technical Report No. 31,* Applied Mathematics and Statistics Laboratory, Stanford University, Stanford, California, October 5, 1956.

McShane, L. M. 1989. Statistical Quality Control Procedures for Monitoring Laboratory Analyses, unpublished Ph.D. dissertation, Cornell University, Field of Statistics, Ithaca, N.Y.

McShane, L. M., and B. W. Turnbull. 1991. Probability Limits on Outgoing Quality for Continuous Sampling Plans, *Technometrics,* Vol. 33, No. 4, November 1991, pp. 393–404.

———. 1992. New Performance Measures for Continuous Sampling Plans Applied to Finite Production Runs, *Journal of Quality Technology,* Vol. 24, No. 3, July 1992, pp. 153–161.

Milligan, G. W. 1991. Is Sampling Really Dead? *Quality Progress,* Vol. 24, No. 4, April 1991, pp. 77–81.

MIL-STD-105E. 1989. *Sampling Procedures and Tables for Inspection by Attributes,* United States Department of Defense, Washington, D.C., 10 May 1989.

MIL-STD-1235A-1. 1979. *Single and Multi-Level Continuous Sampling Procedures and Tables for Inspection by Attributes,* United States Department of Defense, Washington, D.C., 28 June 1974 (Notice 1, 10 March 1979) (Superseded by MIL-STD-1235C Appendices).

MIL-STD-1235C. 1988. *Single and Multi-Level Continuous Sampling Procedures and Tables for Inspection by Attributes,* United States Department of Defense, Washington, D.C., 15 March 1988.

Mundel, A. B. 1991. STANDARDS COLUMN, NATO: Allied Quality Assurance Publications, *Quality Engineering,* Vol. 3, No. 3, pp. 395–404.

Murphy, R. B. 1958. A Criterion to Limit Inspection Effort of Continuous Sampling Plans, *The Bell System Technical Journal,* Vol. 37, No. 1, January 1958, pp. 115–134.

———. 1959a. Stopping Rules With CSP-1 Sampling Inspection Plans in Continuous Production, *Industrial Quality Control,* Vol. 16, No. 5, November 1959, pp. 10–16.

———. 1959b. A Graphical Method of Determining a CSP-1 Sampling Inspection Plan, *Industrial Quality Control,* Vol. 16, No. 6, December 1959, pp. 20–21.

NAVORD OSTD-81. 1952. *Sampling Procedures and Tables for Inspection by Attributes on a Moving Line,* U.S. Navy Bureau of Ordnance, Washington, D.C., August 1952.

Ohta, H., and S. Kase. 1984. GERT Analysis of the Economical Design of Dodge's CSP-1 Continuous Sampling Plan Under Inspection Error, contributing chapter to: *Frontiers in Statistical Quality Control,* Second Edition, Edited by H.-J. Lenz, G. B. Wetherill, and P.-Th Wilrich, Physica-Verlag, Würzburg, Germany.

Okano, F., and W. Wolman. 1956. Operating Characteristics of Continuous Sampling Plans, *Technical Memorandum,* Department of the Navy, Quality Control Division, Bureau of Ordnance, April 1956.

ORDM 608-11. 1954. *Procedures and Tables for Continuous Sampling by Attributes,* Ordnance Corps, Joliet, Illinois, August 1954.

Page, E. S. 1954. Continuous Inspection Schemes, *Biometrika,* Vol. 41, pp. 100–115.

Pesotchinsky, L. 1987. Plans for Very Low Fraction Nonconforming, *Journal of Quality Technology,* Vol. 19, No. 4, October 1987, pp. 191–196.

Phillips, M. J. 1969. A Survey of Sampling Procedures for Continuous Production, *Journal of the Royal Statistical Society—Series A, General,* Vol. 132, No. 2, pp. 205–228.

Prairie, R. P., and W. J. Zimmer. 1970. Continuous Sampling Plans Based on Cumulative Sums, *Journal of the Royal Statistical Society—Series C, Applied Statistics,* Vol. 19, No. 3, pp. 222–230.

QSTAG 340. 1974. *Single and Multi-Level Continuous Sampling Procedures and Tables for Inspection by Attributes,* U.S. Department of the Army, Washington, D.C.

Rajarshi, M. B., and V. S. S. Kumar. 1983. Dodge's CSP-1 for Markov-Dependent Process Using Probability Sampling, *Journal of the Indian Statistical Association,* Vol. 21, pp. 99–111.

Read, D. R., and D. W. Beattie. 1961. The Variable Lot-Size Acceptance Sampling Plan for Continuous Production, *Journal of the Royal Statistical Society—Series C, Applied Statistics,* Vol. 10, No. 3, November 1961, pp. 147–156.

Resnikoff, G. J. 1956. Some Modifications of the Lieberman-Solomon Multi-Level Continuous Sampling Plan, MLP, *Technical Report No. 26,* Applied Mathematics and Statistics Laboratory, Stanford University, Stanford, California, February 8, 1956.

———. 1960. Minimum Average Fraction Inspected for a Continuous Sampling Plan, *Journal of Industrial Engineering,* Vol. 11, No. 3, May-June 1960, pp. 208–209.

Roberts, S. W. 1965. States of Markov Chains for Evaluating Continuous Sampling Plans, *Transactions of the 17th Annual All Day Conference on Quality Control,* Metropolitan Section, ASQC, and Rutgers University, New Brunswick, N.J., September 11, 1965, pp. 106–111.

Rödder, W. 1984. Cost-Optimal Adaptive Quality Assurance in Continuous Production, *QZ (Qualität und Zuverlässigheit),* Vol. 29, No. 7, July 1984, pp. 232–236.

Romig, H. G. 1953. New Statistical Approaches in Aviation and Allied Fields, *Proceedings, Aircraft Quality Control Conference.*

Rosenblatt, H. M. 1954. Naval Ordnance Standard 81, *A Purchasers Continuous Sampling Plan for Attributes Inspection,* Department of the Navy, Quality Control Division, Statistics Branch, Statistical Methodology Section, Washington, D.C.

Rosenblatt, H. M., and H. Weingarten. 1952a. Sampling Procedures and Tables for Inspection on a Moving Line, Continuous Sampling Plans, *NAVORD-OSTD 81,* Bureau of Ordnance, U.S. Navy.

———. 1952b. *The Operating Characteristic Curves for NAVORD OSTD 81,* Department of the Navy, Quality Control Division, Statistics Branch, Statistical Methodology Section, Washington, D.C.

Sackrowitz, H. 1972. Alternative Multi-Level Continuous Sampling Plans, *Technometrics,* Vol. 14, No. 3, August 1972, pp. 645–652.

———. 1975. A Note on Unrestricted AOQL's, *Journal of Quality Technology,* Vol. 7, No. 2, April 1975, pp. 77–80.

———. 1981. ARL Comparisons for Multi-Level Sampling Plans, contributing chapter to: *Frontiers in Statistical Quality Control,* First Edition, Edited by H.-J. Lenz, G. B. Wetherill, and P.-Th Wilrich, Physica-Verlag, Würzburg, Germany.

Satterthwaite, F. E. 1949. A New Continuous Sampling Inspection Plan Based on an Analysis of Costs, *Report No. 130,* Product Service Division, Appliance and Merchandise Department, General Electric Company, Bridgeport, Conn.

Savage, I. R. 1955. A Three Decision Continuous Sampling Plan for Attributes, *Technical Report No. 20,* Applied Mathematics and Statistics Laboratory, Stanford University, Stanford, California, January 22, 1955.

———. 1956. Statistical Production Models and Continuous Sampling Plans, *Technical Report No. 29,* Applied Mathematics and Statistics Laboratory, Stanford University, Stanford, California, August 8, 1956.

———. 1959. A Production Model and Continuous Sampling Plan, *Journal of the American Statistical Association,* Vol. 54, March 1959, pp. 231–247.

Schilling, E. G. 1982. *Acceptance Sampling in Quality Control,* Marcel Dekker, New York, N.Y.

———. 1990–91. Acceptance Control in a Modern Quality Program, *Quality Engineering,* 3(2), pp. 181–191.

———. 1991. Product Oriented Quality Control and Assurance, *Proceedings of the 35th EOQ Annual Conference,* Prague, Czechoslovakia, 17-21 June 1991, pp. 376–382.

Shahani, A. K. 1975. Acceptance Sampling for Continuous Production, *Quality Assurance,* Vol. 1, No. 1, March 1975, pp. 13–16.

———. 1979. Wald-Wolfowitz Type Sampling Plans for Continuous Production, *Technometrics,* Vol. 21, No. 1, February 1979, pp. 21–31.

Sharp, M. C., S. R. Miller, and J. E. Economou. 1989. A General Continuous Sampling Procedure for Resource-Limited Manufacturing Facilities, *Journal of Quality Technology,* Vol. 21, No. 3, July 1989, pp. 163–173.

Sheesley, J. H. 1975. A Computer Program to Evaluate Dodge's Continuous Sampling Plans, *Journal of Quality Technology,* Vol. 7, No. 1, January 1975, pp. 43–45.

Smith, P. 1959. Two Examples of Statistics at Work in Accounting, *Transactions of the 11th Annual All Day Conference on Quality Control,* Metropolitan Section, ASQC, and Rutgers University, New Brunswick, N.J., pp. 177–190.

Sower, V. E., J. Motwani, and M. J. Savoie. 1993. Are Acceptance Sampling and SPC Complementary or Incompatible? *Quality Progress,* September 1993, pp. 85–89.

Stephens, K. S. 1958. A Multi-Level AOQL Sampling Plan with Weighted Defect Classification, *Transactions of the 1958 Rutgers All Day Conference on Quality Control,* New Brunswick, September 1958 and *Western Electric Engineer,* Vol. VI, No. 4, October 1962.

———. 1976. *Quality and Quality Control,* Productivity Series No. 11, Asian Productivity Organization (APO), Tokyo. (Available from APO, 4-14 Akasaka 8-chome, Minato-ku, Tokyo 107, Japan.)

———. 1979. How to Perform Continuous, Skip-Lot and Chain Sampling, *Annual Technical Conference Transactions,* American Society for Quality Control, Milwaukee, Wisc., pp. 207–208.

———. 1981. CSP-1 for Consumer Protection, *Journal of Quality Technology,* Vol. 13, No. 4, October 1981, pp. 249–253.

———. 1995. How to Perform Skip-Lot and Chain Sampling, Second Edition, *The ASQC Basic References in Quality Control: Statistical Techniques,* Vol. 4, American Society for Quality Control, Milwaukee, Wisc.

Stephens, K. S., and K. E. Larsen. 1967. An Evaluation of the MIL-STD-105D System of Sampling Plans, *Industrial Quality Control,* Vol. 23, No. 7, January 1967.

Storer, R. L. 1953a. Tentative Standard for X-Ray Inspection by Continuous Sampling, *ESM 402-0-1,* Ordnance Ammunition Center, Inspection Division, Quality Control Branch, Statistics Section, February 11, 1953.

———. 1953b. Technical Memorandum on the CSP Given in ESM 402-0-1, *ESM 401-0-5,* Ordnance Ammunition Center, Inspection Division, Quality Control Branch, Statistics Section, February 11, 1953.

———. 1953c. Technical Memorandum on Tentative Standard Continuous Sampling Inspection Plans, *ESM 401-0-6,* Ordnance Ammunition Center, Inspection Division, Quality Control Branch, Joliet, Illinois, June 5, 1953.

———. 1956a. The Use of Continuous Sampling in Ammunition Procurement, *Industrial Quality Control,* Vol. 13, No. 11, May 1956, pp. 48–54. (Also published in *ASQC Annual Convention Transactions,* 1954, pp. 523–534.)

———. 1956b. Continuous Sampling Inspection, *Middle Atlantic Conference Transactions,* ASQC, pp. 13–21.

Taylor, W. A. 1994. Acceptance Sampling in the 90's, *ASQC 48th Annual Quality Congress Proceedings,* American Society for Quality Control, Milwaukee, Wisc., pp. 591–598.

Vogt, H. 1986. Application of the Minimax-Regret Principle to Continuous Sampling, *Statistische Hefte,* Vol. 27, pp. 279–296.

Wadsworth, H. M., K. S. Stephens, and A. B. Godfrey. 1986. *Modern Methods for Quality Control and Improvement,* John Wiley & Sons, New York.

Wald, A., and J. Wolfowitz. 1945. Sampling Inspection Plans for Continuous Production Which Insure a Prescribed Limit on the Outgoing Quality, *The Annals of Mathematical Statistics,* Vol. 16, No. 1, March 1945, pp. 30–49.

Wang, S. C. 1976. Process Control . . . a Sample Plan, *Quality,* Vol. 15, No. 12, December 1976, pp. 12–13.

White, J. S. 1961. A New Graph for Determining CSP-1 Sampling Plans, *Industrial Quality Control,* Vol. 17, No. 1, May 1961, pp. 18–19.

———. 1962. Continuous Sampling Plans, *Annual Conference Transactions,* ASQC, May 1962, pp. 87–94.

White, L. S. 1964. On Finding the AOQL's of a Large Class of Continuous Sampling Plans, *Technical Report No. 22,* Statistical Engineering Group, Columbia University, New York, N.Y., May 1, 1964.

———. 1965. Markovian Decision Models for the Evaluation of a Large Class of Continuous Sampling Inspection Plans, *The Annals of Mathematical Statistics,* Vol. 36, No. 5, October 1965, pp. 1408–1420.

———. 1966. The Evaluation of H106 Continuous Sampling Plans Under the Assumption of Worst Conditions, *Journal of the American Statistical Association,* Vol. 61, September 1966, pp. 833–841.

Yang, G. L. 1983. A Renewal-Process Approach to Continuous Sampling Plans, *Technometrics,* Vol. 25, No. 1, February 1983, pp. 59–67.

———. 1985. Application of Renewal Theory to Continuous Sampling Plans, *Naval Research Logistics Quarterly,* Vol. 32, No. 1, February 1985, pp. 45–51.

Zimmer, W. J., and Tai Chiu-Yung. 1980. A General Continuous Sampling Plan for Controlling the AOQ, *Journal of Quality Technology,* Vol. 12, No. 4, October 1980, pp. 191–195.

INDEX

Aasheim, G. L., 59
Abraham, F. L., 22, 29, 59, 63
Acceptable quality level. See AQL
Acceptance number, 1, 6, 14, 22–23, 39
Acceptance sampling, 1, 6
  elements of, 1–2
Acceptance sampling plan, 1
  classification of, 2–3
  elements of, 1–2
  evaluation of, 2
  preparation/development of, 4–6
  types of, 3–4
AFI. See average fraction inspected
Albrecht, L., 59
AMC Manual No. 74-23, 17, 59
Anscombe, F. J., 59
ANSI/ASQC Standard A2-1987, 1, 2, 3, 59
ANSI/ASQC Standard S1-1987, 4, 55
AOQ, 27
  curves for, 34, 35, 36, 37, 38
  equations for, 12, 16, 24, 44, 45, 47, 49–50
AOQL
  achieving, 8
  AQL and, 17, 19, 26, 34, 35, 36, 37
  as sampling base, 5, 27, 28, 29, 30
  clearance interval and, 19
  defect allowance and, 14
  determination of, 27, 30, 31, 32
  LQL and, 26, 34
  nomographs and, 24
  probability limits on, 9
  protection of, 2–3
  sampling inspection and, 8, 9
  size of, 28, 29
  unrestricted, 9
AQAP-3, 59
AQL
  AOQL and. See AOQL
  as sampling base, 5
  producer risk, 2
Arp, D. L., 59
Attribute inspection, 13, 22, 59, 61, 62, 63, 64, 65
Automatic inspection, 13
Average fraction inspected (AFI), 12, 24, 34, 35, 37, 38
Average outgoing quality. See AOQ

Average Outgoing Quality Limit.
  See AOQL

Banks, J., 12, 59
Banzhaf, R. A., 9, 13, 18, 22, 59
Barron, C. L., 59, 60
Bauer, P., 60
Beattie, D. W., 60, 65
Bennett, E. G., 9, 60
Biedenbender, R. E., 17, 60, 63
Binomial probability distribution, 2
Blackwell, M. T. R., 9, 19, 60
Block sampling, 8
Bowker, A. H., 17, 60, 64
Breeze, J. D., 60
Brugger, R. M., 9, 13, 18, 22, 59, 60
Bulk inspection, 5

Case, K. E., 9, 49, 60
Chain sampling, 1, 2, 4, 66
Chiu, W. K., 9, 60
Chiu-Yung, Tai, 67
Clearing interval, 7–8, 13, 19, 37, 38, 39
Connolly, C., 60
Consumer, 3, 5, 6, 17, 34, 35, 36, 65
Consumer risk, 3
Continuous inspection, 5, 7
Continuous Sampling Plans, 4. See also CSP
Control chart, 6
Corrective action, 17, 34, 36
Criteria for conformity, 5
CSP, 7
CSP-A, 38
  defect processing and, 13
  features of, 17
  MIL-STD-1235A and, 18
  sampling tables for, 34, 35–36, 51
  schematic for, 18
CSP-C, 22–24, 39
CSP-F, 19, 34, 37, 39
CSP-M, 39
  CSP-V and, 39
  features of, 17–19
  MIL-STD-1235A and, 18
  schematic for, 20–21
  sampling tables for, 34, 36–37, 53–58
CSP-1, 38
  application of, 13

best plan of, 29–30, 34
CSP-C and, 22, 23
CSP-F and, 19
CSP-T and, 37
evaluation of, 9, 12–13
extensions of. *See* CSP-2; CSP-3
measures of performance, 12
nomographs for, 24, 25, 27, 28, 29, 30, 31
OC curves for, 9, 11, 34
operating characteristics for, 41–44, 45
operation of, 9
parameters of, 7–9
probability of acceptance in, 26, 44
replacement contraction in, 12, 49–50
schematic for, 9
tables for, 34
CSP-R, 22, 29
CSP-T, 19, 22, 34, 37, 39
CSP-2, 38
    AOQL and, 26, 27
    CSP-C and, 23
    CSP-T and, 37
    evaluation of, 14–16
    nomograph for, 24, 25, 32
    nonreplacement and, 29
    OC curves for, 35
    operating characteristics for, 41, 42–43, 44–45
    schematic for, 14
    tables for, 32
CSP-3, 14, 23, 26, 28, 38
    evaluation of, 16
    operating characteristics for, 41, 42–43, 46–47
    rule of four and. *See* rule of four
    schematic for, 15
    tables for, 34
CSP-V, 19, 22, 23, 34, 37–38, 39
Cumulative-Results Sampling, 4
Curram, J. B., 60

Dau, D. E., 59, 60
Davison, W. R., 59, 60
Defects
    classification of, 5
    clearing interval for, 13
    critical, 37
    occurrence of, 7, 14
    percentage of, 2, 7, 9, 24, 25, 26, 36
    removal of, 9, 12, 23, 49
    spacing of, 7, 8
    stopping rule and, 17
Derman, C. S., 8, 9, 17, 19, 60, 61

Destructive test, 5
Detailing, 1
Discrete attributes, 3
Dodge, H. F., 61
    as father of acceptance sampling, 6
    continuous vs. lot-by-lot sampling, 8
    CSP-N, 17
    CSP-1, 7, 13, 47, 49
    CSP-2, 14
    CSP-3, 14, 16, 26, 47
    developer of control chart, 6
    nomographs and, 24, 25
    random order of defect spacing, 8
    replacement vs. nonreplacement, 12
    skip-lot sampling, 28
    spotty quality, 9
    tables for sampling plans by, 4
Double Sampling Plans, 4
Duncan, A. J., 2, 61

Economic design, 9
Economou, J. E., 66
Elfving, G., 17, 61
End-of-line inspection, 13
Endres, A., 9, 61
Environmental test, 5
Equilibrium probability, 16, 41, 44, 46
ESM 401-0-6a, 61
ESM 401-0-17, 61
Ewan, W. D., 62

Fibonnacci series, 26
Finite length production runs, 9, 19, 37
Fordice, J. J., 18, 62
FORTRAN programs, 12, 47

General continuous sampling plan (GCS Plan), 22
Gessford, J., 17, 62
Ghare, P. M., 62
Girshick, M. A., 62
Glossary and Tables for Statistical Quality Control, 2
Godfrey, A. B., 5, 6, 62, 66
Govindaraju, K., 22, 23, 24, 26, 34, 62, 63
Grant, E. L., 2, 62
Gregory, G., 9, 62
Guide, H., 59
Gurfel, B., 62
Guthrie, D., 17, 29, 62

Hackl, P., 60
Hansen, B. L., 62

Hassan, M. Zia, 18, 62
Heldt, J. J., 60, 62
Hiatt, W. R., 60
Hillier, F. S., 9, 29, 62
H-106, 17, 18, 62
H-107, 17, 62

Incoming quality, 9
Indifference quality, 2
In-line inspection, 13
In-process inspection, 5
Inspection, 1, 4, 5
  error in, 9
Ireson, W. G., 17, 63
ISO TR 8550:1994, 5

Jafri, S. Q., 63
Johns, M. V., Jr., 8, 9, 17, 29, 61, 62
Johnson, N. L., 59
Journal of Quality Technology, 13
Juran, J. M., 63

Kandasamy, C., 22, 23, 24, 63
Kao, E. P. C., 63
Kase, S., 64
Kelly, H. W., 22, 63
Kemp, K. W., 62
Koopman, L., 63
Kosik, P., 63
Kotz, S., 59
Kumar, V. S. S., 9, 18, 63, 65

Lack of control, 9
Larsen, K. E., 4, 66
Lasater, H. A., 9, 63
Leavenworth, R. S., 2, 62
LeMaster, V., 9, 63
Lenz, H.-J., 60, 63, 64, 65
Lieberman, G. J., 8, 9, 17, 61, 63–64
Life test, 5
Limiting quality (LQ), 3
Limiting Quality Level. See LQL
Littauer, S., 17, 19, 60
LQL, 2, 3, 5, 25, 26, 34
Lot-by-lot sampling, 1, 4, 8, 9, 12, 14, 26
Lot inspection, 5, 7, 13
Lot size, 1, 5, 6
LTPD, 3

MacLean, A., 59
Magwire, C., 9, 17, 64
Markov chain, 16, 23, 41
McKeague, R., 9, 63

McShane, L. M., 9, 12, 64
Measures of performance, 12, 13, 23–24
Miller, S. R., 6, 66
Milligan, G. W., 64
MIL-STD-105, 54, 57
MIL-STD-105E, 3, 4, 22
MIL-STD-123 (ORD), 19
MIL-STD-1234 (ORD), 34
MIL-STD-1235 (ORD), 17, 18, 35, 36, 51
MIL-STD-1235A, 17, 22, 34
MIL-STD-1235B, 17, 34
MIL-STD-1235A-1, 64
MIL-STD-1235C, 39, 64
  CSP-A and, 17
  CSP-F and, 19, 37
  CSP-1 and, 34
  CSP-T and, 19, 37
  CSP-2 and, 35
  CSP-V and, 19, 22, 38
  LQL counterpart in, 3
  sampling tables and, 34
Minimum average fraction inspected, 9
Motwani, J., 7, 66
Multilevel plans, 6, 18, 19, 36, 37
Multiple Sampling Plans, 4
Multistage subsampling, 6
Mundel, A. B., 5, 62, 64
Murphy, R. B., 8, 9, 29, 30, 33, 64

NAVORD OSTD-81, 17, 64
Nomographs, 24–33
  for "best" CSP-1 plan, 29, 33
  for CSP-1, CSP-2, 25, 27, 29, 30, 31, 32
  with LQL Sections, 28
Nonconformance, 2
Non-discrete variables, 3
Nonreplacement case, 12, 23, 29, 34, 49–50
Normal inspection, 6, 22, 36

OC curve
  as evaluation tool, 2
  continuous sampling and, 9
  CSP-A and, 17, 36
  CSP-F and, 37
  CSP-1 and, 13, 34
  CSP-2 and, 35
  CSP-V and, 38
  example of, 3
  lot-by-lot sampling and, 9
Ohta, H., 64
Okano, F., 17, 64

On-line inspection, 5
Operating Characteristic Curve. *See* OC curve
Operation schematic
  for CSP-A, 18
  for CSP-C, 24
  for CSP-M, 20–21
  for CSP-1, 11
  for CSP-T, 22
  for CSP-3, 15
  for CSP-2, 14
  for CSP-V, 23
ORDM 608-11, 64

Page, E. S., 64
Pesotchinsky, L., 64
Phillips, M. J., 64
Prairie, R. P., 64
Probability of acceptance
  continuous sampling and 9, 24
  CSP-1 and, 26, 44
  CSP-2 and, 45
  CSP-3 and, 47
  during sampling, 25
  lot-by-lot sampling and, 9
  OC curve and, 2, 3
Probability sampling, 8, 17, 47
Process quality, 6
Process fraction defective, 12
Producer, 2, 6, 29, 30, 36
Producer's nominal quality level (PNQL), 29
Producer's risk, 2
Production interval, 17, 35, 36, 38
Product unit, 5
P_t scale, 26

QSTAG 340, 17, 64
Quality characteristics, 5
Quality control, 6
Quality models, 9
Quality Engineering, 7

Rajarshi, M. B., 9, 63, 65
Random order plan, 7
Random sampling, 8
Read, C. B., 59
Read, D. W., 65
Reduced inspection
  attribute sampling and, 22
  CSP-A and, 17, 35–36
  CSP-M and, 18–19, 36
  CSP-V and, 19, 37, 39
  when used, 6

Rejectable quality level (RQL), 3
Rejection, 1, 4; 6
Repairing, 1
Replacement case, 12, 34, 49–50
Reprocessing, 1
Resnikoff, G. J., 9, 17, 29, 65
Response characteristics, 2
Roberts, S. W., 41, 65
Robustness, 9
Rödder, W., 65
Romig, H. G., 4, 61, 65
Rosenblatt, H. M., 65
Rubin, H., 62
Rule of four, 16, 18, 41, 43, 46

Sackrowitz, H., 9, 18, 65
Sample size, 4, 5, 6, 8
Sampling, 1, 6–7. *See also* acceptance
  sampling plan
Sampling tables, 4, 34–38, 51, 53–58
Sarkadi, K., 63
Satterthwaite, F. E., 9, 65
Savage, I. R., 9, 65
Savoie, M. J., 7, 66
Schematics. *See* Operation schematic
Schilling, E. G., 6, 65
Schmidt, J. W., 9, 60
Scrapping, 1
Screening
  acceptance and, 1
  AOQL and, 29–30
  check of, 19, 34, 37, 38
  resubmissions and, 17
  state of, 17
  verification of, 17, 19, 22, 35, 36
  work space for, 13
Shahani, A. K., 65, 66
Sharp, M. C., 66
Sheesley, J. H., 12, 47, 66
Skip-lot inspection, 1, 2, 4, 6, 28, 66
Skip-lot sampling, 1, 4, 13, 28
Smith, P., 66
Solomon, H., 17, 19, 60
Sorting, 1
Sower, V. E., 7, 66
Specification limits, 5
Spotty quality, 9, 14, 18, 25
Stephens, K. S., 1, 4, 5, 6, 13, 16, 28, 66
Stopping rule
  as evaluation measure, 9
  basis of, 17

CSP-F and, 37, 39
CSP-M and, 36
CSP-1 and, 34, 39
CSP-2 and, 35, 39
CSP-3 and, 39
CSP-T and, 37, 39
CSP-V and, 38, 39
when activated, 19, 22
Storer, R. L., 59, 60, 66
Systematic sampling, 8

Taylor, W. A., 7, 66
Test method, 5
Thompson, P., 59
Tightened inspection, 6, 17, 19, 22, 35, 36, 39
Torrey, M. N., 7, 8, 9, 14, 16, 24, 26, 47, 61
Total quality control, 6
Total quality management (TQM), 6
Transition probability matrix, 41, 43
Turnbull, B. W., 9, 12, 64

Type 1 error, 2
Type 2 error, 2
Unit Sequential Sampling Plans, 4

Variables data, 3, 4
Verification sampling, 17, 19, 22, 35, 36
Vogt, H., 66

Wadsworth, H. M., 6, 66
Wald, A., 66
Wang, S. C., 66
Warning rule, 19. *See also* stopping rule
Weingarten, H., 65
Wetherill, G. B., 9, 60, 63, 64, 65
White, J. S., 9, 18, 29, 67
Wilrich, P.-Th., 60, 63, 64, 65
Wolfowitz, J., 66
Wolman, W., 17, 64

Yang, G. L., 67

Zimmer, W. J., 64, 67